Ezra Rhead

Metallurgy

An elementary text-book

Ezra Rhead

Metallurgy
An elementary text-book

ISBN/EAN: 9783337275976

Printed in Europe, USA, Canada, Australia, Japan

Cover: Foto ©berggeist007 / pixelio.de

More available books at **www.hansebooks.com**

METALLURGY

AN ELEMENTARY TEXT-BOOK

BY

E. L. RHEAD

LECTURER ON METALLURGY AT THE MUNICIPAL TECHNICAL SCHOOL
MANCHESTER

LONDON
LONGMANS, GREEN, & CO.
AND NEW YORK
1895

PREFACE

In issuing this little work the author has endeavoured to provide the student with a book of moderate size, giving a clear and concise account of metallurgical processes, and the principles upon which they are based. Details are only given when necessary for the sake of clearness.

The chemical changes involved in the various processes described are specially dealt with, but it must be remembered that the equations given for reactions occurring at elevated temperatures in most cases only partially express the truth.

The book is intended as a text-book for students commencing the study of Metallurgy, and as a small handy book of reference. It has been written in as popular a style as the subject permits, to make it available for the general reader.

The author wishes to express his indebtedness to Messrs. Fraser and Chalmers for several of the diagrams in the chapters on Gold and Silver; and to his former students, Mr. W. H. Mortimore, for Figs. 51 and 55; Mr. Jno. Allan and Mr. W. McD. Malt for assistance in correcting the later proofs.

MANCHESTER,
August, 1895.

CONTENTS

INTRODUCTION.

CHAPTER I.

PHYSICAL PROPERTIES OF METALS.

CHAPTER II.

METALLURGICAL TERMS AND PROCESSES.

CHAPTER III.

FURNACE TYPES.

CHAPTER IV.

REFRACTORY MATERIALS.

CHAPTER V.

FUELS.

CHAPTER VI.

GASEOUS FUELS.

CHAPTER VII.

IRON, ETC.

CHAPTER VIII.

IRON SMELTING.

CHAPTER IX.

CHEMICAL REACTIONS OCCURRING IN, AND PRODUCTS OF THE BLAST FURNACE.

CHAPTER XIV.

MERCURY.

CHAPTER XV.

SILVER.

CHAPTER XVI.

GOLD.

CHAPTER XVII.

TIN.

CHAPTER XVIII.

ZINC.

CHAPTER XIX.

OTHER METALS.

CHAPTER XX.

ALLOYS.

THE science of Metallurgy deals with the properties of metals in the different conditions they may assume—the changes in these properties induced by the treatment to which they are subjected, or brought about by the influence of other substances with which they may be mixed, either as impurities or for some useful purpose. It also treats of the methods by which they are extracted, in a more or less pure state, from the substances in which they occur naturally, and the refining of the crude products.

The properties on which the usefulness of a metal depends are: specific gravity, hardness, toughness, tenacity, elasticity, malleability, ductility, expansion by heat, fusibility, resistance to atmospheric and chemical action, conductivity for electricity and heat, and the manner in which it affects the properties of metals with which it may be mixed.

The high specific gravity of gold reduces coins of considerable value to a reasonable size, while the low specific gravity of iron, compared with its strength, reduces the weight of iron structures. A gold structure of the same strength as an iron one would be nearly nine times as heavy.[1]

The hardness of steel fits it for cutting-instruments. The toughness, malleability, ductility, and tenacity determine the

[1] $$\frac{\text{Tenacity of gold}}{\text{Specific gravity}} = \frac{7}{19 \cdot 6} = 0 \cdot 362$$

$$\frac{\text{Tenacity of iron}}{\text{Specific gravity}} = \frac{25}{7 \cdot 8} = 3 \cdot 205$$

$$\frac{3 \cdot 205}{0 \cdot 362} = 8 \cdot 853 \;\ldots$$

workability and general usefulness of a metal for structural and constructive purposes; the fusibility and expansibility its suitableness for making castings; while a greater or less resistance to atmospheric corrosion is necessary for its general application.

Useful Metals.—Of the fifty-five elements classed as metals by the chemist, only some twenty occur in such quantity, or possess properties which raise them to such importance, as to be of consequence to the metallurgist. These are—

Iron	Lead	Platinum	Chromium	Cobalt
Copper	Antimony	Aluminium	Mercury	Cadmium
Zinc	Gold	Manganese	Magnesium	Sodium
Tin	Silver	Bismuth	Nickel	Potassium

CHAPTER I.

PHYSICAL PROPERTIES OF METAL.

Specific Gravity, or comparative density, is the weight of the metal, bulk for bulk, compared with water. It is generally increased by mechanical treatment, such as hammering, rolling, and wire-drawing.

TABLE OF SPECIFIC GRAVITIES.

Water = 1.

Magnesium	1·74	Nickel	8·8
Aluminium	2·56	Bismuth	9·2
Antimony	6·7	Silver	10·5
Zinc	7·1	Lead	11·36
Tin	7·2	Mercury	13·0
Iron	7·8	Gold	19·3
Copper	8·6	Platinum	21·5

HARDNESS.—This property is very much affected by the purity of the metal and the treatment which it has undergone. Speaking generally, the hardness of a metal—with few exceptions—is increased by the presence of impurities. Gold for coinage is hardened by 8·33 per cent. of copper, and the presence of a small percentage of carbon in iron converts it into steel. Other examples will be found in the text. By suitable treatment, steel may be made hard enough to scratch glass, or soft enough to be turned and worked freely. (See Tempering Steel.) Mechanical treatment, such as hammering, wire-drawing, rolling, and pressure in the cold state, hardens metals. In this manner the bronze weapons of the ancients were hardened. Heating to redness, and allowing to cool very slowly, generally has the effect of softening the metal. In the case of copper this is reversed, rapid cooling, such as quenching in water, softening that metal and its alloys. Metals are usually softer when hot than when cold.

Fracture is the appearance presented by metals when ruptured.

Metallic fractures may be classified as—

Crystalline.—Metals presenting this appearance are weak, as fracture occurs by the separation of the adherent facets. Antimony, bismuth, and zinc offer good examples of this kind of fracture.

Granular.—This fracture presents the structure of a sandstone. The homogeneity of the mass is greater than when crystalline, and the metal is consequently stronger and more readily worked. Cast iron is a good example of this structure.

Fibrous.—This structure is developed to the greatest degree in wrought iron by the elongation and welding together of the particles during rolling. The toughness and strength of this metal are too well known to require comment.

Silky.—This is a finely fibrous fracture of brilliant silky lustre. It is best seen in copper. Metals which possess it are strong, tough, and malleable.

Conchoidal.—This appearance is presented by the harder varieties of steel. The metal breaks with a convex or concave surface with divergent markings somewhat resembling a shell. Metals possessing this fracture are hard, highly elastic, and brittle.

Columnar.—The columnar structure is manifested by the tendency of the metal to separate in long fingers across the thickness of the cake or ingot, the pieces somewhat resembling lump starch. It is obtained by heating the metal nearly to its melting-point, and then either allowing it to fall on the ground, or by striking it sharply with a wooden mallet while hot. Tin and lead are the best examples of this structure.

The fracture of a metal varies with the purity, temperature, and manner in which the rupture has been produced: *e.g.* wrought iron containing phosphorus breaks with a crystalline fracture; copper at a full red heat breaks with a coarsely granular fracture; while wrought iron, if nicked *all round* and broken short off, may present a granular fracture, but if nicked on one side only, and then bent over and broken, it exhibits a fibrous fracture.

Fusibility.—All metals, with the exception of chromium, have been reduced to a fluid condition by heat. The readiness with which this can be done and the degree of heat required to effect it vary greatly. Tin, lead, and zinc melt in an ordinary fire, platinum only in the oxy-hydrogen blowpipe flame. Many metals, before fusing, pass through a soft, pasty stage, *e.g.* iron and platinum; others pass directly from the solid to the liquid state. This applies to alloys also. The alloy of two parts lead and one part tin, used by plumbers to make the joints of lead pipes, is an excellent instance of this, the knob of metal round the pipe being wiped on and shaped with the metal in the pasty state.

Some metals contract on fusion, and are denser in the fluid than the solid state at the point of melting. For this reason, solid lumps of cast iron will float in fluid metal of the same kind till melted. Most metals expand on liquefying, and are denser in the solid state. They go through exactly the reverse changes on cooling.

Metals which expand on solidifying bring out finer impressions of moulds when used for castings. Certain qualities of iron are consequently superior for this purpose, and for a similar reason bismuth is added to zinc and tin alloys, for casting the so-called artificial "bronzes," and for making alloys for patterns.

When metals are mixed together to form alloys, the melting-point of the mixture is lowered, sometimes in a remarkable degree, even below the melting-point of the most fusible constituent, *e.g.* a mixture of 1 lead, 1 tin, 2 bismuth melts in boiling water. (For melting-points of tin and lead alloys, see Alloys.) This is taken advantage of in the production of the so-called "fusible alloys," which are required to melt at a certain temperature, and for solders. Alloys used for this purpose must melt more readily than the objects to be soldered.

The fluidity of metals when melted is very variable. For casting purposes, the metal must flow freely, or portions of the mould will not be filled—cold shorts—and the sharpness will be destroyed.

TABLE OF MELTING-POINTS.

Tin	230° C.	Silver	950° C.
Bismuth	268°	Copper	1050°
Lead	330°	Gold	1075°
Zinc	412°	Cast Iron	1200–1300°
Antimony	450°	Iron	about 1600°
Aluminium	700°	Platinum	„ 1770°
Magnesium	800°		

Volatility.—Some metals are readily converted into vapour by heat, and are described as volatile metals. Such metals can be distilled, the vapour being led into condensers and cooled. Mercury, zinc, cadmium, sodium, potassium, and arsenic are obtained from their ores in this way, the vapour of the reduced metal being led away from the reduction chambers, or retorts, and condensed.

NOTE.—Volatility is only a relative quantity. Almost all metals are volatilized to a greater or less extent at very high temperatures, such as are obtainable in the electric arc and furnace, while lead, antimony, gold, and silver, are sensibly volatile at furnace temperatures.

Tenacity.—Resistance to fracture by a stretching force is possessed by all metals in a greater or less degree. It is expressed by the amount of dead weight which a bar of given sectional area can support without rupture. In English measures it is expressed as the number of pounds or tons supported by a bar one square inch in section; in metric measure, as kilogrammes per square millimetre.

TABLE OF RELATIVE TENACITIES.

Steel	100	Gold	12
Wrought Iron	30–40	Zinc	2
Cast Iron	10–24	Tin	1–3
Wrought Copper	18–30	Bismuth	1–1·5
Cast Copper	12–25	Lead	1
Cast Silver	25	Lead (wire)	1·5–2·5
Aluminium	20–28	Antimony	0·8

The steel taken as 100 has a tenacity of 60 tons per square inch.

This property is greatly affected by the purity of the metal. The presence of *certain* impurities in *some* cases increases it, while in others the tenacity is diminished by foreign matters. The presence of the small amount of carbon in iron necessary to convert it into steel is attended by a marked increase in tenacity. The presence of silicon, on the other hand, diminishes it. Many other cases will be found in the sequel. Excess, even of a salutary kind of impurity, often lowers the tensile strength, as is the case with the larger proportion of carbon present in cast iron. Metal which has been mechanically treated, as by hammering or rolling (especially in the cold), or by wire-drawing, is generally stronger than cast specimens of the same metal. Thus, steel wire, No. 14 gauge, 0·087 inch in diameter, drawn from steel rods having a tenacity of fifty-seven tons, has a tensile strength of ninety-eight tons.

Mechanical treatment seems to produce some change in structure, especially in the external parts. In a wire, a hard skin is formed on the surface, the proportion of which to the whole bulk varies with the gauge of the wire. If this be immersed in acid and dissolved off, the interior is found to possess little or no greater tenacity than the original metal.

This increase in strength is reduced to its ordinary level by heating the metal to full redness and allowing it to cool slowly, *i.e.* **annealing**.

Heat, if excessive, lowers the tensile strength. The degree of heat varies with different metals. The tenacity of metals often changes, as the result of the situation in which it is employed. Iron and steel frequently become crystalline and brittle by continued vibration, or by frequent heating to redness and cooling, and are in consequence weakened. Many fractures result from this cause.

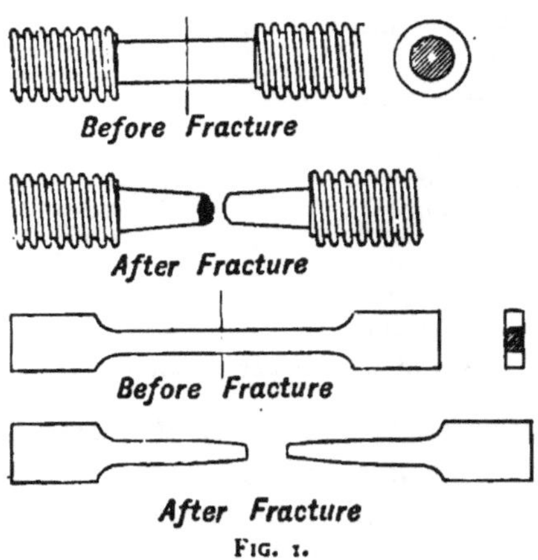

Fig. 1.

Tenacity is determined by straining a piece of metal of known dimensions, and observing the amount of force necessary to fracture it.

Fig. 1 shows forms of test-pieces for various purposes before and after fracture.

The ends are securely gripped and force applied.

The force is generally applied by means of hydraulic pressure acting on a ram to which one of the grips (shackles) is attached, and the orce expended is weighed, by either a simple or compound lever arrangement, much on the same principle as a common steelyard.

Figs. 2 and 3 show diagrams of simple and compound lever arrangements.

Testing-machines for determining tenacity are generally provided with appliances for other purposes.

Sometimes, instead of weighing the force, the pressure employed is registered by gauges, and the force calculated.

The force required to fracture a piece is generally greater if applied at once than when gradually applied.

Elasticity is the amount of force which can be resisted without permanent deformation or "set" being produced. It will be observed from Fig. 1 that the pieces after testing are longer than before. If during the test the strain is relieved from time to time by removing the force, it will be found that the piece assumes its *original* length until a certain amount has been exceeded. After that the test-piece becomes permanently elongated. Up to that point the substance is perfectly elastic, and the amount of force required to produce permanent lengthening marks its "limit of elasticity." The proportion

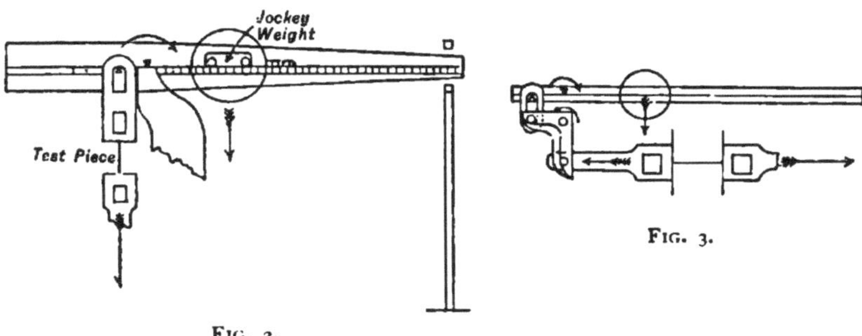

FIG. 2.

FIG. 3.

which this bears to the tenacity of the body is of importance in structural work. The larger the proportion, the more reliable will be the material, and the less likelihood of its being affected by vibration, etc.

The "**modulus of elasticity**" is the force that would be required to double the length of a bar if its elasticity remained perfect. The "modulus" is an index of the stretching capacity of the metal.

Elongation.—The extent to which a metal elongates prior to fracture is a matter of the greatest importance. Tough metals usually show a considerable increase in length. Hard, brittle metals elongate but little.

Important evidence as to the working qualities of the material is thus furnished. To determine the elongation, the test-piece is measured between the points at which it is gripped before and after straining till fractured, and the increase stated in percentage of the original length. Thus a 10-inch test-piece of boiler steel measured 12·5 inches after fracture, *i.e.* 2·5 inches over 10 inches = 25 per cent. Elongation is accompanied by a diminution in area of section. This is measured in order to determine whether the elongation was local or uniformly distributed. Sometimes the contraction in area is confined to the region of fracture. Results are thus stated—

Description of Sample of Mild Steel.

Tensile strength in tons per square inch.	Elastic limit.	Elongation per cent.	Contraction of area.
28	15	25	40

Curves are often drawn automatically or plotted from results, showing the behaviour of the piece at different loads.

Ductility is the property which permits of the body being drawn out in the direction of its length—that is, converted into wire. The metal from which the finest wire is producible is the most ductile. Wires are produced by dragging rods of a convenient size through holes in a steel-faced plate, somewhat smaller than the rod itself, and repeating the operation till it is reduced to the desired gauge. The hole is slightly tapered, and the end of the rod is ground down sufficient to permit of its being thrust through the hole far enough to grip it tightly. The metal becomes hard and brittle by this treatment, and requires to be frequently annealed. The ductility is much less hot than cold, so that all wires are drawn cold. The property is dependent principally upon the tenacity, and in a less degree upon the hardness. Metals which are moderately soft and fairly tenacious are the most ductile. The tenacity must be great enough to resist the force necessary to pull it through the holes, this being dependent upon the hardness.

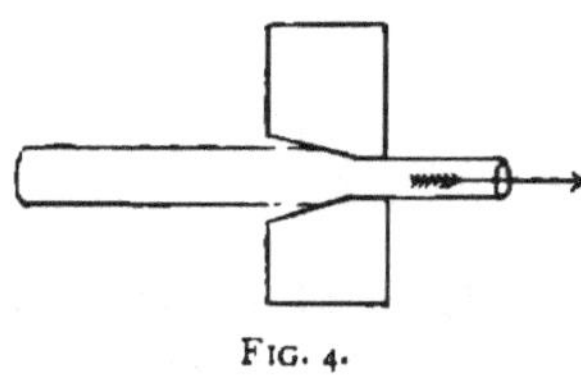

Fig. 4.

On this account gold and silver head the list of ductile metals, and iron precedes copper, tin, lead, etc., whose tenacity is much inferior, although they are softer than iron.

Wire-drawing in most cases increases the density, the force employed being converted into pressure by the conical form of the hole through which it is drawn.

Order of Ductility.

Gold	Aluminium	Zinc
Silver	Iron	Tin
Platinum	Copper	Lead

Gold wires as fine as the threads of a spider's web have

been drawn by enclosing the gold in silver, and dissolving off the latter in nitric acid after drawing down.

Malleability.—Metals which can be beaten out or otherwise extended in all directions are said to be malleable, the degree being measured by the thinness of the leaves it is possible to produce. The effect of hardness upon this property is much more pronounced than on the previous one. It is owing to this cause that copper stands so high, and iron so low, in the scale of malleability, while in point of ductility iron exceeds copper. When the pressure necessary to overcome the hardness and spread the metal is greater than the tenacity, rupture occurs. Malleability is seriously affected by the presence of impurities, in some cases, a trace of certain substances being sufficient to destroy it. Traces of bismuth, arsenic, or antimony in gold make the metal brittle. It is also greatly affected by temperature. Iron is most malleable while hot, but must not be overheated, or it becomes burnt. Commercial zinc affords a striking instance of the effect of heat in this respect. In the cold, the metal is brittle and crystalline. At a temperature of 120° C. to 150° C. it is malleable, and can be rolled into thin sheets; at a somewhat higher temperature it becomes more brittle than when cold. Sheets of zinc rolled at the proper temperature retain a considerable degree of malleability, and can be bent and worked like other sheet metals with a little care.

ORDER OF MALLEABILITY.

Gold	Copper	Lead
Silver	Tin	Zinc
Aluminium	Platinum	Iron

Plates, sheets, foil, and leaf, are terms applied to different thicknesses of metal.

Plates, sheets, and foil are generally rolled. Leaf is beaten out by hammering. Gold leaf $\frac{1}{280000}$ of an inch in thickness, and so thin as to transmit light, is commonly produced by hammering. At the Great Exhibition of 1862, sheets of Russian iron were shown $\frac{1}{700}$ of an inch thick. These, it would seem, had been produced by hammering the sheets in packets with charcoal powder between. Various tests are applied to determine malleability, such as bending, hammering, etc. Such articles as rivets and angle irons should be very malleable.

Toughness is the resistance which the metal offers to fracture by bending or twisting.

Most malleable metals are tough, but not always in proportion to their malleability. It is determined by the number of times the metal can be bent to and fro before breaking, or the number of twists that can be put on a wire or rod of given length.

In some cases, as in testing steel rails, a heavy weight is allowed to drop from a given height upon the rail resting on supports.

Purity is not always associated with the extreme of toughness. Tin, for example, renders copper tougher; while phosphorus in iron renders it "short"—in the cold, "cold short;" and sulphur makes it "short" when hot—"red short."

Cold short metal works fairly well above a red heat, and red short below that temperature. The term "short," as here applied, means lack of toughness and malleability.

Flowing Power.—Metals which in the *solid* state can be readily shaped into any required form by pressure are said to possess the flowing property. Stampings, lead pipes and rod, coins, medals, etc., are made by taking advantage of this property. It does not in any way refer to the fluidity of the metal when molten, the operations being conducted upon the metal in the solid state.

The property seems to depend upon a combination of malleability, ductility, and toughness, allied to a peculiar structure giving the metal a semi-plastic character something like half-dried glaziers' putty, which permits of the particles rolling over each other freely. Lead possesses the power of "flowing" to a great extent. In consequence of this, the plumber is able to work up by gentle hammering lead vessels from sheet; the superfluous metal being gradually worked away into the sides, making them thicker.

Lead pipe is squirted from a solid ingot by means of hydraulic pressure, the tube being formed by a mandril or die, as it passes from the press.

In striking medals and coins, a disc of metal (a blank) is placed between steel dies, and the sudden application of pressure causes it to flow and fill all the finest lines of the die. The great sharpness of medals and coins is due to this method of production. If cast, the metal would solidify before completely filling the mould owing to its thinness.

Welding.—If two pieces of certain metals are pressed together under suitable conditions, they unite and form one piece *without the use of any solder.* This is called welding. It is essential that the surfaces in contact shall be perfectly clean, and that they shall be in such condition as *to flow* readily under pressure. Most metals require heating to a greater or less extent before the flowing-point is reached. Lead welds in the cold or when only slightly heated. Iron, on the other hand, has to be heated almost to whiteness. Most metals oxidize when heated, and, to ensure a good weld, it is necessary to remove the oxide in order to secure perfect contact between the uniting surfaces. It is, in consequence, difficult to weld lead, as it oxidizes superficially in moist air, if only exposed for a short period at ordinary temperatures. Gold, on the other hand, if pure, unites with the greatest ease, being sufficiently soft and also unoxidizable.

In welding iron, the metal is either made hot enough to fuse the oxide formed, or else sand is used, which, by combining with it, forms a fusible body (silicate of iron). In either case, when the pieces are placed together, and the junction hammered, the fluid matter is squeezed out (hammer-slag), and chemically clean faces come into contact. Borax is often used instead of sand for the same purpose.

Lead welds to tin without much difficulty. If a clean sheet of lead is overlaid with a sheet of tin, and passed through rolls, they unite, a compound sheet resulting.

The metals which weld readily are platinum, gold, silver, lead, tin, iron, and nickel.

In electric welding, the ends to be united are placed together, and a powerful electric current of low tension is passed by suitable connections from one piece across the point of contact to the other. The high resistance at the junction, owing to the poor contact, causes the development of intense local heat, which is greatest at the faces to be joined. When hot enough, the ends are forced together by a screw arrangement, and union between the pieces takes place. (Thomson's Process.)

In welding large iron tubes made from plate, rings, etc., the electric arc is employed, the arc being sprung between the tube itself, suitably supported, and carbon rods manipulated by hand, or otherwise suspended above it. (Bernado's Process.)

Conductivity.—Metals are, speaking generally, good conductors both for heat and electricity. Their relative conducting powers are as follows :—

	For heat.[1]	For electricity.[2]
Silver	1000	1000
Copper	748	941
Gold	548	730
Aluminium	—	511
Zinc	—	266
Platinum	94	166
Iron	101	155
Nickel	—	120
Tin	154	114
Lead	79	76
Bismuth	18	11

Electrical conductivity is greatly diminished by a rise in temperature and by impurities. Impure copper may have a conductivity little superior to that of iron. Alloys as a rule are poor conductors, but are less affected by heat.

CHAPTER II.

METALLURGICAL TERMS AND PROCESSES.

COMPARATIVELY few metals are found to any great extent in the metallic condition. When occurring in that form they are said to be **native.** The whole of the platinum, and practically all the gold used, is thus found. Silver, copper, mercury, bismuth, and arsenic also occur native in notable quantities.

Native metals occur in bodies of considerable size, as threads and filiform masses penetrating the rocks, in grains more or less minute distributed through the rock mass, or in alluvium, in thin flakes, and associated with other substances containing the metal.

N.B.—Masses of native copper 500 tons' weight have been found in the Lake Superior district, and nuggets of gold weighing 183 lbs. in Victoria. Native metals are often crystalline. Alluvium is the *débris* which results from the wearing down of rocks.

Metals generally occur in chemical combination with other elements, whereby their metallic character is completely

[1] Matthieson. [2] Franz and Wiedemann.

masked. When a mineral contains a sufficient quantity of a metal combined with some element from which it can be readily separated, so as to render the extraction of metal of good quality profitable, it is said to be an **ore** of the metal. The most commonly occurring compounds from which the metals are principally obtained, are—

Oxides	Metal and oxygen	Iron, copper, zinc, tin, manganese, chromium, antimony, and aluminium
Sulphides	Metal and Sulphur	Copper, lead, zinc, antimony, silver, mercury, bismuth, cadmium
Carbonates	Metal, carbon, and oxygen	Iron, copper, zinc, lead, manganese
Fluorides	Metal and fluorine	Aluminium
Chlorides	Metal and chlorine	Copper, silver
Phosphates	Metal, phosphorus, and oxygen	Lead
Arsenides	Metal and arsenic	Cobalt and nickel
Silicates	Metal, silica, and oxygen	Copper, zinc, nickel

Hydrated oxides and carbonates contain water. Oxy-chlorides contain metal, oxygen, and chlorine. Bromides, iodides, and many complex substances also occur as ores.

The quantity of metal required to make the working profitable depends on the value of the metal extracted, and the form in which it occurs. A few pennyweights of gold per ton of ore, if in the free state, can be satisfactorily worked; while an iron ore must contain about 20 per cent. of the metal to be profitable.

Iron ores also afford a good instance of the effect of the combination in which the metal exists. Iron pyrites contains 46 per cent. of iron, but it is combined with sulphur, from which element it is difficult to completely separate it, and the iron made from the material, after burning off the sulphur, is of inferior quality owing to the tenacity with which that element is retained. Antimonial gold ores furnish another example.

Ores are sometimes found in deposits following the general lie of the rocks in which they occur. Such deposits are known as **beds**. When the occurrence is irregular, the ore being accumulated at certain points, the deposit is called a **pocket** or **bunch**. Many ores are found in what appear to have been fissures or cracks, which have been filled up with material altogether differing from the rocks in which they occur. These deposits are called **veins** or **lodes**. They do not follow the stratification of the rocks, but cut through them at a greater or less angle. Veins of quartz are often called *reefs*. The line along which they reach the surface is the *outcrop*.

Owing to the action of the air, moisture, etc., the upper part of a vein is often entirely altered, and sometimes has spread over the surface, forming a *cap;* or the alteration may extend deeper, even to the water-line.

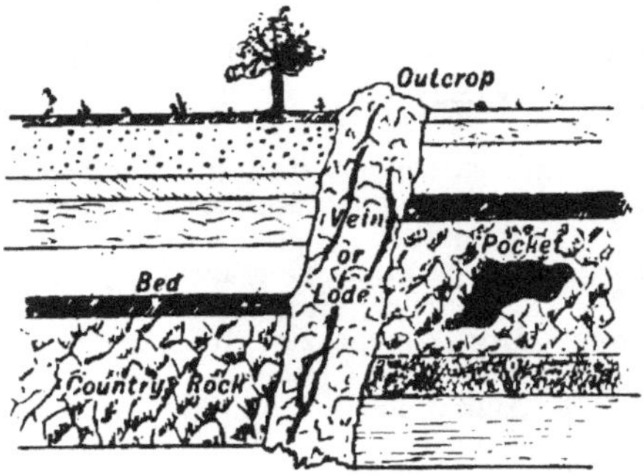

Such alteration may have led to a complete change of the chemical character of the vein, sulphides giving rise to sulphates, oxides and carbonates to hydrated oxides, etc.[1]

Fig. 5.

The rock lying on either side of a vein is the *country* rock.

Veins are filled with various materials, some of which are of a metallic and others of a non-metallic nature, and often lumps of country rock are included. In Fig. 5 the black portions are the ore bodies. The non-metallic portion is known as *veinstuff.* It has generally a lower specific gravity than the metallic portion. The materials commonly found as veinstuffs are quartz, chlorite, felspar, mica, hornblende, and other silicates, barytes, fluor, calcite, dolomite, etc.

The operations necessary to separate these from the metallic portion are called **dressing the ores.**

Much of the ore can often be separated in a sufficient degree of purity by simply picking it over by hand, and breaking away adherent rock with a hammer. This is known as *hand-picking.*

Fig. 6.—Stone-breaker.

When it is mixed up with the veinstuff, more elaborate treatment is necessary. The methods employed are generally based on the different specific gravities of the materials to be separated. The metalliferous portion of an

[1] See iron.

ore is generally heavier than the veinstuff with which it is associated.

Heavy substances settle out more rapidly when agitated with water, and are less easily carried forward by a running stream than lighter ones, and are more quickly deposited.

The ore is first broken up, a **stone-breaker, crushing-rolls,**

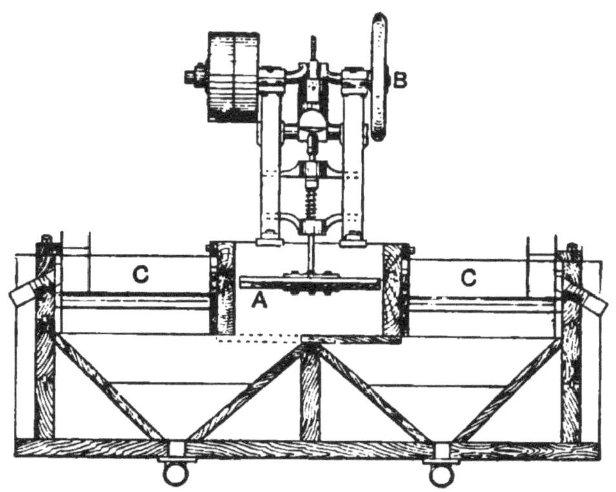

FIG. 7.—Plunger Jig. A, plunger; C, screens; B, driving-gear for plungers.

grinding-mill, or **stamps** being employed, according to the degree of fineness required. The broken stuff is then separated into sizes by a series of screens, the different portions being separately treated.

Material which is not too fine is washed in **jigs**. These consist of sieves or shallow boxes with bottoms of wire cloth, suspended in water, and jerked up and down by mechanical means; or the water is forced upward through the material in jerks by means of a plunger. The disturbance thus produced causes the heavy material to gravitate to the bottom, and the light matters can be scraped or washed off the top.

The fine stuff is dressed by subjecting it to the action of a current of water on sloping tables.

Buddles (Fig. 8) are circular, slightly conical tables, upon which the fine material, suspended in water, is fed at the apex. Water is supplied, and the ore stirred by brushes attached to revolving arms. The light portions are carried away by the water, and the heavy material accumulates on the cone, the heaviest nearest the apex.

FIG. 8.—Buddle.

Racks and Washing-tables are inclined tables on which the material is placed at the higher end and washed down by

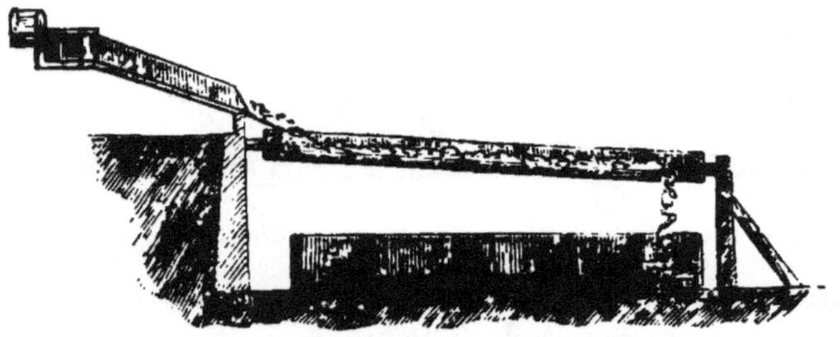

FIG. 9.—Rack used in washing tin ores.

a gentle stream of water, being pushed back against the current by brushes or rakes. The light stuff is washed away.

FIG. 10.—Frue Vanner.

The **Frue Vanner** (Fig. 10), now largely employed, consists of a wide endless belt of indiarubber so stretched on rollers that the top forms an inclined table. A rapid shaking motion is communicated to the table, and the belt slowly travels in an upward direction. The fine stuff is fed with water from the trough at the higher end, and clean water is also sprayed on to the table. The current, aided by the jerking motion, separates and carries off the earthy matters, and the heavy metallic portions are carried on by the belt and washed off in the trough under the frame. Vanners are specially suitable for treating very fine material.

The dressed ore as delivered by the miner to the smelter is still impure. The earthy matters still associated with it are known as **gangue**.

Smelting.—The various operations whereby the metal is separated *by fusion* from the ore are known as *smelting*. The smelting campaign often involves several distinct operations.

Reduction.—The separation of the metal from chemical combination is known as *reduction*. If an oxide, this is generally done by heating it with carbon or carbonaceous matters, such as charcoal, coal, or coke; the carbon of these substances combining with the oxygen and forming CO_2 (carbonic acid gas) or CO (carbon monoxide), according to the temperature at which the reduction occurs. CO itself, is a powerful reducing agent, combining with oxygen, and forming CO_2. Hydrogen reduces oxides, with the formation of water (H_2O).

Sulphides are sometimes reduced directly to the metallic state by heating with iron or with iron-bearing materials. Thus galena (sulphide of lead) yields sulphide of iron and metallic lead—

$$2PbS + Fe_2 = 2FeS + Pb_2$$

and stibnite (sulphide of antimony) yields sulphide of iron and antimony—

$$Sb_2S_3 + 3Fe = 3FeS + Sb_2$$

The substance employed to liberate the metal is the **reducing agent.** In this case the iron unites with the sulphur, and liberates the metal.

Sulphides are often reduced by **air-reduction processes.** Thus cinnabar (sulphide of mercury) is reduced by simple heating in a current of air. The sulphur burns off, leaving the mercury—which is volatilized by the heat—free. The vapour is condensed.

$$HgS + O_2 = SO_2 + Hg$$

Lead and copper sulphides are also reduced by air-reduction processes.

The ore, as received from the mine, is often not suitable for immediate separation of the metal, either because the quantity present is too small, or owing to the combination in which it occurs. Copper pyrites seldom contains sufficient of the metal, and zinc blende contains its zinc as sulphide, whereas the metal is best obtained from the oxide.

The preliminary treatment to which the ores are subjected generally takes the form of heating them in a plentiful supply of air. This process is called **calcination.** By this means the sulphur in sulphides is burnt off as sulphur dioxide, which being gaseous passes away, and the metal at the same time also takes up oxygen from the air, and is converted into an oxide. Or the removal of the sulphur may only be partial, and a sulphate may result—

$$\underset{\text{Lead sulphide}}{PbS} + \underset{\text{Oxygen from air}}{3O} = \underset{\text{Lead oxide}}{PbO} + \underset{\text{Sulphur dioxide}}{SO_2}.$$

$$\text{or, } 2PbS + 7O = PbO + \underset{\text{Lead sulphate}}{PbSO_4} + SO_2$$

The sulphides of iron, copper, lead, silver, and zinc thus form sulphates during calcination. The amount formed depends on the temperature and other conditions of the roasting. With the exception of lead sulphate, they are decomposed by strongly heating them. Iron, copper, and zinc sulphates yield oxides. Silver sulphate is reduced to metal.

Arsenic is similarly removed as white arsenic, As_2O_3 (see Tin Smelting), and antimony to some extent as antimonious oxide, Sb_2O_3. During calcination other changes of great importance take place. Carbonates are decomposed with the expulsion of carbonic acid gas (CO_2), leaving an oxide of the metal.

$$\underset{\text{Zinc carbonate (Calamine)}}{ZnCO_3} = \underset{\text{Zinc oxide}}{ZnO} + \underset{\text{Carbonic acid gas}}{CO_2}$$

Moisture is expelled, and in some cases protoxides, *i.e.* oxides containing the lowest proportion of oxygen, are converted into higher oxides. This is sometimes of great importance, as in iron smelting. The introduction of protoxide of iron into the furnace would seriously impair its working, besides causing loss of iron in the slag. All ores, therefore, containing this oxide must be calcined before introduction to the furnace, when the following change occurs :—

$$3FeCO_3 + O = Fe_3O_4 + 3CO_2$$

Carbonate of protoxide of iron, or ferrous carbonate — Magnetic oxide of iron

Calcination also leaves the material in a more open and porous state, and it is thus more readily acted upon during reduction, especially by gaseous reducing agents such as carbon monoxide.

The term "roasting" is often used in the same sense as "calcining." In copper smelting it only applies to the operation in which the metallic copper is separated.

Most metals are converted into oxides by calcining. Gold, platinum, and silver are not affected.

The calcined material may be at once reduced, or first subjected to a series of operations for the purpose of concentrating the metal in smaller bulk, from which enriched portion it is finally obtained. This is done by taking advantage of some chemical property manifested by the metal being worked for, to separate it from foreign matters.

NOTE.—The greater part of the copper obtained, is produced from copper pyrites, $Cu_2SFe_2S_3$—a compound of iron and copper sulphides—which should contain 34 per cent. of copper. It is usually mixed, however, with so large a proportion of iron pyrites, FeS_2, that it seldom contains more than 12 per cent. of copper, and often less. The concentration of the copper is brought about by taking advantage of the superior affinity of copper for sulphur and of iron for oxygen. By calcining the ore, some iron and copper sulphides are changed to oxides, but on heating the whole to fusion, the copper oxide is decomposed by the unaltered iron sulphide remaining, copper sulphide and iron oxide resulting. The iron oxide is removed by combining with silica, in the slag, the copper sulphide, being heavier, sinking to the bottom of the furnace. The material is thus enriched in copper, and after one or two treatments consists of practically pure sulphide of copper, from which the metal is extracted.

A mixture of sulphides obtained artificially in this manner by fusion is called a **regulus**, or **matte**. Cobalt and nickel are concentrated as arsenides. The mixture of arsenides is called a **speiss.**

Smelting operations are conducted at a temperature above the melting-point of the metal. Most metals, after reduction, are obtained in a fused state, and, being heavier than the other materials, run down and form a lowermost layer in the furnace or crucible. Zinc, mercury, cadmium, sodium, and potassium are vaporized at the temperature of reduction, and the vapours are led away and condensed.

Fluxes.—The infusible earthy matters often present in ores may seriously impede the collection of the reduced metal, or retard the reduction by enveloping it and preventing the action of reducing agents, or, by combining with it chemically at the high temperature, cause loss of metal in the slag. It is therefore necessary in smelting to provide means for causing them to be liquefied at the furnace temperature. This is done by mixing with the ore and reducing agent some substance which either melts itself and dissolves the infusible matter, or, by combining with it in the furnace, forms a substance which is fusible at the temperature employed.

Fluor spar, for example, dissolves barytes and phosphate of lime, and lime combines with clay and forms a fusible body.

Substances added to the furnace charge for this purpose are called *fluxes.*[1]

Most fluxes act to a large extent both chemically and physically. The earthy matters to be removed are divisible into two great classes. Those consisting of earthy metallic oxides and carbonates (the CO_2 is expelled during smelting, and oxides are produced), such as limestone, dolomite, etc., are *basic* in character; silica (quartz, sand, etc.), and many other substances containing it, are known as *acid* gangue. When silica is heated with oxides of metal, combination takes place, and bodies called silicates are produced. Thus, lime and silica form silicate of lime, and so on. Some of these melt readily, others only at the highest temperatures. The fusibility depends on the nature of the metallic oxide, and on the amount present. Silicates of soda and potash, lead, manganous, and ferrous

[1] *Fluo*, to flow.

silicates melt comparatively easily, but silicates of lime, magnesia, alumina, and zinc are practically infusible at ordinary furnace temperatures. When, however, more than one metallic oxide (base) is present in combination with the silica, forming a compound or complex silicate, the mixture of the two silicates is much more readily fusible; the more fusible the silicates employed are separately, the lower will be the temperature at which the mixture will melt. Thus, common soft glass is a mixture of silicate of soda and silicate of lime; flint glass, of silicates of lead and potash. Thus, also, by mixing silicates of lime and alumina or magnesia, fusible bodies are produced.

From the foregoing it will be seen that the selection of a flux will depend on the nature of the gangue to be removed. If *silica* only, then some oxide whose silicate is fusible, as oxide of iron, must be employed, or two bodies such as lime and alumina or magnesia. If clay (silicate of alumina) is to be removed, an addition of lime is all that is necessary. If *metallic oxides, or basic bodies*, have to be fluxed, silica must be added, and, if necessary, a second metallic oxide to produce a fusible body.

The substance produced by the combination of the flux with the gangue is called a **slag** or **cinder**. In most cases they are mixtures of silicates, and thus partake of the chemical nature of glass. Their appearance depends very much on the rate of cooling, and their composition. Rapid cooling gives a glassy, and slow cooling a stony, appearance. When gases escape during solidification, the slag is full of holes—vesicular or spongy.

The principal materials employed as fluxes are—

Substance.	Character.	Composition.
Lime	Basic . .	CaO
Limestone	,, . .	$CaCO_3$
Mountain limestone . .	,, . .	$CaCO_3MgCO_3$
Alumina	,, . .	Al_2O_3
Clay	Acid . .	Al_2O_3 and SiO_2, etc.
Quartz, sand, etc. . . .	,, . .	SiO_2
Oxide of iron and slags containing it	Basic . .	Fe_2O_3 and Fe_3O_4
Fluor spar	—— . .	CaF_2

Garnet, felspar, and other natural silicates are sometimes employed. Borax, and carbonate and sulphate of soda, are also used in small quantities in special operations. Borax is sodium biborate, and dissolves metallic oxides, forming fusible borates. At high temperatures, the soda it contains combines with silica, and thus acts as a flux for that substance. The use of fluor as a flux for barytes and phosphate of lime (bone ash) has already been referred to. Fluor is also a flux for silica. When strongly heated together, a *gaseous* fluoride of silicon is formed and lime remains, which is fluxed off in the usual manner.

$$2CaF_2 + SiO_2 = SiF_4 + 2CaO$$

The *bases* generally found in slags are lime, magnesia, alumina, ferrous oxide (FeO), manganous oxide, and smaller quantities of potash and soda.

NOTE.—The ferric and magnetic oxides of iron do not readily combine with silica, but when heated with reducing agents, ferrous oxide is formed, which is a powerful flux.

In many refining processes, slags are produced containing the metal under treatment. These are subsequently worked up for the recovery of the metal.

Most silicates are capable, when fused, of carrying in suspension or solution excess either of the metallic oxide present or of silica. If the metallic oxide is in excess, the slag is said to be *basic;* if silica, it is described as *acid* or *siliceous.* Silicates are generally classified according to the ratio existing between the oxygen in combination with the metal and silicon respectively.

Sub-silicates	$4RO.SiO_2$	$4R_2O_3.3SiO_2$	2 : 1
Mono- ,,	$2RO.SiO_2$	$2R_2O_3.3SiO_2$	1 : 1
Sesqui- ,,	$4RO.3SiO_2$	$4Al_2O_3.9SiO_2$	1 : 1½
Bi- ,,	$RO.SiO_2$	$R_2O_3.3SiO_2$	1 : 2
Tri- ,,	$2RO.3SiO_2$	$2R_2O_3.9SiO_2$	1 : 3

A slag is *clean* when the metal has been so completely extracted as to permit of its being thrown away. An ore is said to be *self-fluxing* or *self-going* when the earthy constituents are fusible without the employment of a flux. When a mass of materials is fused, the substances formed separate according to their relative specific gravities, and the slag, being lightest, floats on the top. Sometimes metal, speiss, regulus, and slag are produced in the same operation. They arrange themselves in the order stated.

Refining Processes.—Metals when first obtained are never pure, and many methods are followed for their purification. The refining process adopted depends on the metal under treatment and the impurities to be removed. In some cases, as with iron and antimony, the reducing agents employed, carbon and iron respectively, are taken up by the metal to a certain extent. They are eliminated by heating the metal first obtained with more of the ore, oxide of iron in the case of iron (puddling), and sulphide of antimony with antimony. In each case further reduction takes place at the expense of the foreign matters present.

In most cases, however, the impurity consists of foreign metals present in the ore, and simultaneously reduced, together with sulphur, arsenic, etc.

These are generally removed by exposing the metal at a high temperature or in a molten state, to the oxidizing influence of the air in a suitable furnace. The oxides which form are removed from the surface by skimming, or, if the heat is sufficient, unite with silica and form fusible silicates. The method of conducting the operation, and the name it receives, differ with the metal under treatment.

Lead is thus *improved*, iron *refined*, copper *scorified.*

The term **scorification** (L., *scoriæ* = ashes) is also applied to a process in the dry assay of gold, silver, and other ores. A quantity of the silver ore is mixed with finely granulated metallic lead, placed in a clay dish (scorifier), and heated in a muffle until *about half the lead is oxidized.* The silver and gold are set free, and alloy with the residual lead.

It is sometimes possible to separate the greater part of the impurity by carefully melting out the metal from the less fusible impurity. This process is called **liquation**, and the term applies generally to the separation of matters according to their different melting-points, spontaneously, as during solidification, or otherwise. Alloys are frequently not homogeneous from this cause.

In refining tin, the impure metal is placed on a sloping bed and gently heated. The tin melts first, liquates out, and drains away, leaving the bulk of the impurities at the upper end of the hearth. Antimony sulphide is also separated by liquation from the infusible matters with which it is associated in the ore. (See also Lead Refining.)

Silver is purified by alloying it with lead and then removing the lead by oxidation. This is conducted on a *cupel* made of bone ash or marl brasque (see Silver), and the process is known as **cupellation.** The oxide of lead (litharge) formed is fusible, and is either run off the surface or partially absorbed by the bone-ash bed. The silver and gold being unoxidizable are unaffected, and remain on the cupel. Base metals present are attacked by the oxide of lead, and the oxides formed, although not fusible at the temperature at which the process is carried on, are dissolved by the molten litharge and carried off, leaving the precious metals pure.

The separation of silver and other metals from gold is the object of the operation known as **parting.** This process consists of dissolving out the silver by the action of acid, leaving the gold unattacked.

CHAPTER III.

FURNACE TYPES.

Most metallurgical operations are conducted in structures specially designed either for the production and employment of high temperatures, or to secure perfect control of the temperature and gaseous atmosphere in which the process is carried on. In many cases, special features are introduced with a view to saving fuel.

Classification.

(1) *Kilns and Stalls.*—Structures or enclosures in which the materials are mixed with the fuel, free access of air is permitted, and no fusion takes place.

(2) *Hearths.*—Shallow and more or less open fireplaces, in which the materials and fuel are mixed, a blast of air supplied, and the atmosphere made more or less oxidizing by varying the amount of air supplied.

(3) *Wind Furnaces.*—Deep fireplaces, with grates at bottom and flue openings at top, for heating crucibles, etc. (Fig. 50).

(4) *Blast Furnaces.*—Tall structures in which the materials and fuel are mixed together, an air-blast introduced near the bottom, and in which fusion of the contents is effected.

(5) *Reverberatory Furnaces.*—Furnaces in which the fuel is burnt in a separate part of the chamber, the flame and hot gases only, coming into contact with the material treated.

(6) *Muffle Furnaces.*—Chambers which are heated by the flame, etc., circulating in flues which surround them.

(7) *Tube and Retort Furnaces.*—Furnaces in which the operation is conducted in vessels fixed in a chamber and heated.

(8) *Regenerative Furnaces.*—Those in which the waste heat is employed for heating the air, or air and gas, supplied to the furnaces.

Kilns.—Calcining operations are frequently conducted in vertical chambers provided at the bottom with a grate or with openings to admit air. The substance to be calcined is mixed with sufficient fuel, the burning of which generates the heat necessary to carry on the operation. Gjer's calciner for calcining iron ore is shown on p. 91.

Kilns are sometimes heated by gas or by the waste heat from furnaces.

Reguli and mattes are often calcined in stalls (Fig. 11) usually built in blocks, back to back. The back wall contains the main flue, which communicates by the openings O, and by flues in the side wall with the interior of the stall. The front is loosely built up, and the top covered with small stuff and a sheet of corrugated iron while the operation is going on. With reguli rich

FIG. 11.

in sulphur, a good layer of wood at the bottom, to start the operation, is all the fuel required, the heat generated by the burning sulphur, etc., being sufficient to carry it on. Several calcinations are needed to completely remove the sulphur, a larger proportion of fuel being required each time. Coke or coal is frequently used in the later stages.

Fig. 12 represents a **blast furnace**. On examining the figure it will be noticed that the vertical furnace chamber has no grate, the bottom of the furnace being masonry or other solid material. An air-blast is supplied to the furnace by means of bellows, fans, blowers, or blowing-engines, through nozzles, which enter at D. These nozzles are called *tuyeres*. The materials to be treated are charged into the furnace along with the fuel, and remain in contact with it throughout. As the substances melt, they run down to the bottom and accumulate in the space below the tuyeres, known as the crucible, or hearth. When sufficient has collected, an opening (kept stopped with clay) is made into the furnace, and the melted matters allowed to flow out, or they may flow out continuously into a separate receiver. It is obvious that in such furnaces fusions and processes of a reducing character only can be conducted, since the materials are heated in contact with carbonaceous bodies employed as fuel.

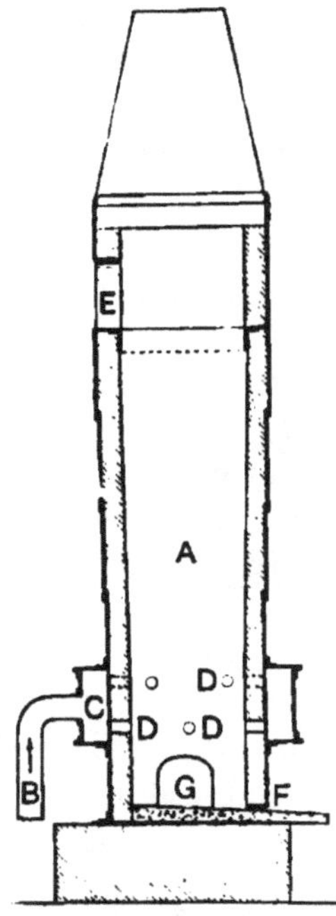

FIG. 12.—Foundry Cupola for melting Iron. A, stack; B, blast-pipe; C, blast-box; D, tuyeres; E, charging opening; F, tap-hole; G, cleaning door.

Economy of Fuel in Kilns and Blast Furnaces.—In calcining in kilns, the air admitted at the bottom finds its way up through the descending hot material, and cools it, thus carrying the heat back into the kiln, while the descending column of cold material charged in at the top deprives the ascending current of hot gases (products of combustion, etc.), of much of their sensible heat, carrying it thus downwards into the kiln. For maximum economy of fuel, the combustion should take place in the middle region.

In *blast furnaces*, the combustion takes place near the region at which air is blown in, and the ascending stream of gases is cooled by the material in the upper part of the furnace, the degree of cooling depending on the rate of ascent and the height of the column. From the blast furnaces used in smelting iron, they escape at a temperature of from 200° C. to 300° C. As the temperature of combustion is high, the carbon burns to CO. Any attempt to burn this by blowing in air higher up the furnace is met with the difficulty of establishing a second region of combustion.

Fig. 13 represents a type known as the **Reverberatory Furnace.**

It will be seen that the chamber in this case is horizontal, and is divided into two unequal parts by a low partition (fire-bridge) crossing it. The smaller part is the *fireplace,* closed with fire-bars below, and with an opening for charging the fuel. The larger portion is the *laboratory* of the furnace, the bottom

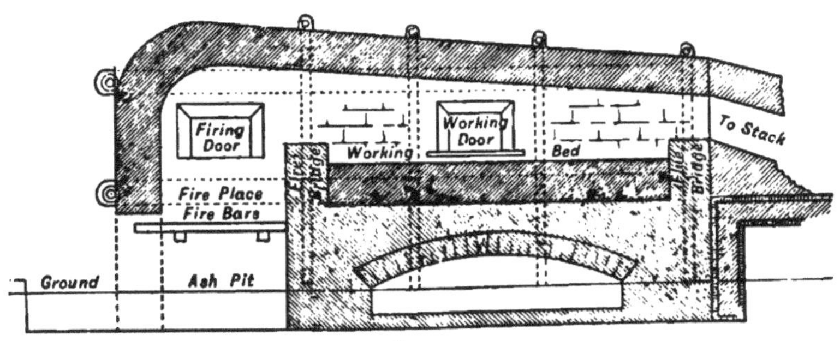

FIG. 13.

of which is the *bed* or *hearth,* and on this the materials are treated. Flues at the end opposite to the fireplace communicate with the stack or chimney. The roof gradually inclines towards the flues, and reflects (reverberates) the flame and hot gases from the fire downward, and, getting heated, radiates heat on to the bed. In furnaces of this class, it will be observed, the materials and fuel do not come into contact, and hence all kinds of operations can be conducted in them. Thus, by mixing reducing agents with the charge, reduction can be effected (see Tin and Lead Smelting), and by admitting air to the furnace-chamber, through openings in or near the fire-bridge, the substances under treatment are heated in contact with air, and oxidation (calcination) goes on. Sometimes air is blown in, as in cupellation, etc. By regulating the air-supply to the fire with dampers, the atmosphere can be made reducing or oxidizing as may be required.

The draught is sometimes aided by forcing air through the fire by a steam-jet injector similar to that shown in Fig. 74. The steam issuing at high pressure from the nozzle at the mouth of the trumpet-shaped tube, entangles and carries forward the air. Not much steam is required, only about 5 per cent., at a pressure of 60 lbs. The horizontal branch passes under the fire-bars, and the ash-pit is closed by doors luted round with clay.

Fig. 15 shows what is known as a water-jacketed furnace. In these furnaces those parts subjected to the most intense heat and the action of corrosive slags, etc., are made of hollow iron casings through which water circulates, the cooling action of which prevents the iron from being affected.

Muffle Furnaces.—In some cases it is necessary, for various reasons, to exclude the products of combustion as well as the

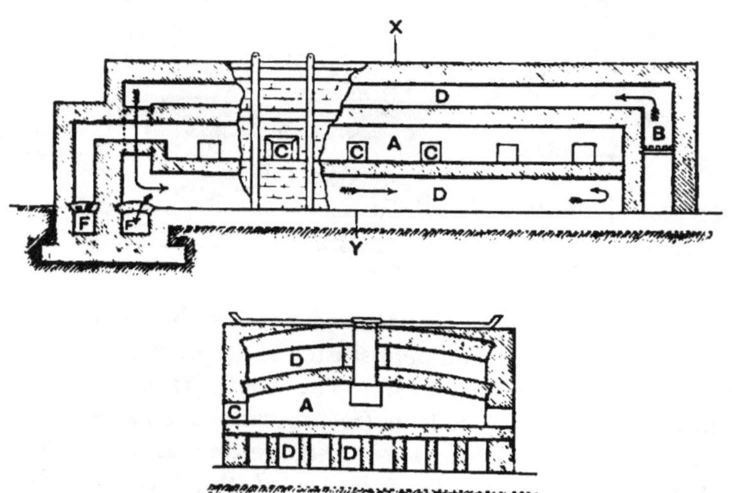

Fig. 14.—Muffle Furnace. A, chamber; B, fireplace; C, doors; D, flues round chamber; F, flues to stack, etc.

fuel. In such circumstances *muffle furnaces* are employed. The muffle is a chamber surrounded by the fire, or by flues through which the products of combustion and hot gases from the fire pass. Such a furnace as used in copper extraction is shown in Fig 14. Muffle furnaces are also used in assaying silver and gold.

In **Regenerative Furnaces**, the heat carried away to the flues by the escaping gases, is employed to heat the air

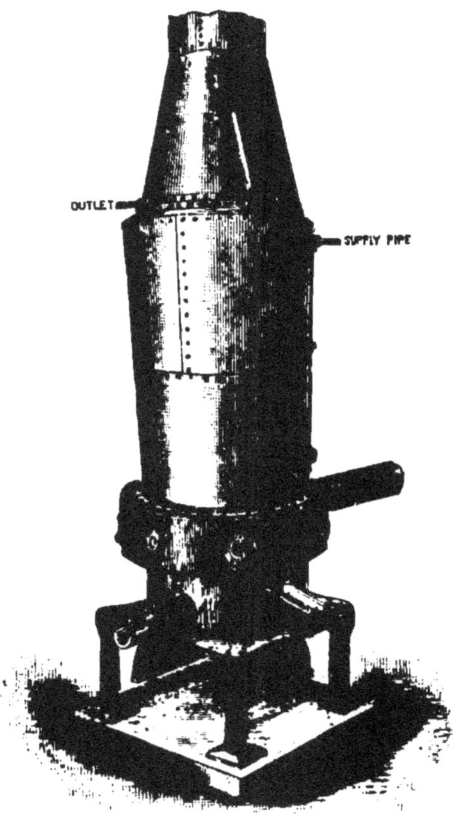

FIG. 15.

supplied to the fire, and thus returned to the furnace, effecting a considerable saving in fuel. In gas-fired furnaces, the gas is also heated before burning. Siemens's regenerative furnace is described on p. 154.

Tube and Retort Furnaces consist merely of a fire-chamber in which retorts or tubes for the reception of the materials to be treated are suitably supported. They are employed in the extraction of bismuth, zinc, etc. (see p. 236).

Mechanical furnaces of various forms designed to effect mechanically what in ordinary furnaces is done by hand are extensively employed. In calcining fine materials, the continual or repeated turning over of the material so as to expose it to the air, is necessary, and involves much manual labour. In the *Brückner* furnace (Fig. 16), this is done by putting the material into a brick-lined chamber, as shown, which can be caused

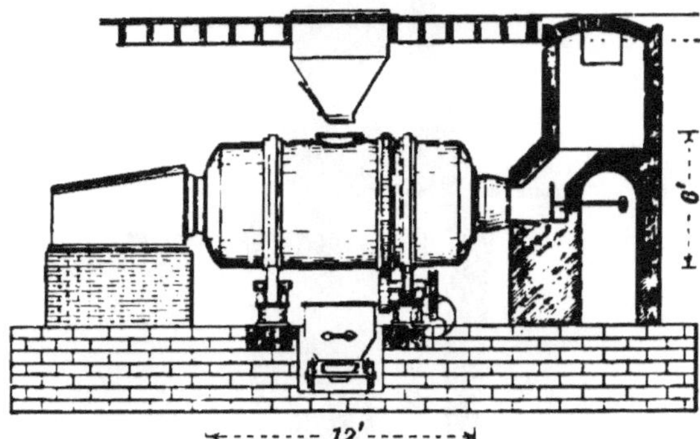

FIG. 16.—Brückner Calciner.

to revolve slowly. The chamber is carried on rollers, and the motion communicated by the gearing shown. It makes about six revolutions per minute. The fireplace is stationary, and the flue is provided with a damper. In the White-Howell furnace (Fig. 17), the revolving chamber is placed at a small angle; the ore, fed continuously from a hopper at the higher end, is gradually moved forward by the rotation of the chamber, being

FIG. 17.—White-Howell Furnace.

picked up by projections inside, and dropped again as the furnace revolves. The roasted matter is discharged at the lower end.

In Gerstenhoffer's calciner, the finely divided ore is fed on to triangular shelves crossing the furnace, and arranged so that each row of shelves catches what trickles from those above, and thus exposes it fully to the hot air and gases from the fire.

In tower furnaces, shown in Fig. 18, the fine material is allowed to fall down tall heated chambers, and meets in its descent an ascending current of hot gases from the fires F, and air admitted through suitable openings. The sulphur and other combustible bodies are oxidized, and the gases escape by the flue B. The door at C is for the removal of the roasted material, and DD to remove the dust carried over by the current of gases.

In other forms of calciner, rakes and ploughs are caused to periodically

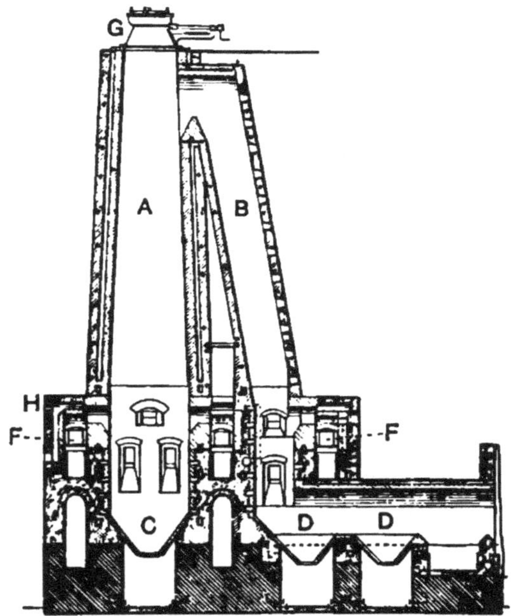

Fig. 18.—Stetevelt Furnace. A, tower; B, descending flue; C, discharge door; D, dust-hoppers; F, fireplaces; G, feed hopper.

traverse the bed or beds, and turn over the material to expose fresh surfaces; or, as in Brunton's calciner, the bed is horizontal, and revolves. Projections from the roof turn the material over, and gradually move it towards the edge, where it is discharged.

The structure of a furnace may be divided into two parts: that portion which gives support and stability, and the portion specially adapted to resist the heat and the action of fluxes and slags. The latter constitutes the *lining* of the furnace chamber. The outer supporting part generally consists of common brickwork or masonry, often strengthened by iron bands, and tied together by transverse rods, supported by iron plates (buck plates, from "buckle," "to bend"), and strengthened at intervals by thick iron plates or flanges (buck staves). These are fastened together by means of iron rods across the furnace—tie-rods—to prevent accident from the expansion and contraction of the masonry. The outer masonry should be badly conducting material.

CHAPTER IV.

REFRACTORY MATERIALS.

THE substances employed for lining furnaces are required to withstand high temperatures and the corrosive action of such substances as they come into contact with in the furnace, and to possess in certain cases other important characters.

Fire-clay.—The most important and most generally used material is *fire-clay*. These clays consist mainly of hydrated silicate of alumina, $Al_2O_32SiO_22H_2O$ (alumina, silica, and water), with an excess of silica, and are marked by the small amounts of lime, magnesia, oxide of iron, potash, and soda which they contain. From the remarks on fluxes (see p. 19), the effect of these substances on clay, in producing fusibility, will be at once apparent.

No silicate of alumina is quite fusible at furnace temperatures, and when excess of alumina or silica is present the body is even more refractory. Analyses of various clays will be found on p. 33.

The water of hydration present is in chemical combination, and cannot be removed by drying at the boiling-point of water. Its presence in the clay gives to it one of its most important properties, viz. that of taking up water mechanically mixed with it, and becoming soft and plastic. Clay does not take up the maximum amount at once, but only gradually, so that previous to use clay is tempered with water and mellowed by exposure. The water taken up mechanically *can* be removed by drying.

When clay is burned, the water of hydration is expelled,

and a hard *anhydrous* substance remains. This body has no power of taking up water and becoming plastic, and no artificial means are known of restoring the clay to its original state. The expulsion of the water during burning causes clay to contract, and allowance has to be made for this. In bricks, blocks, slabs, and other articles of simple form, this is done by merely making the dimensions of the body just large enough to allow for the contraction.

This, however, cannot be done in the case of crucibles, retorts, and other fire-clay ware. Owing to the unequal contraction of parts of different thickness, they would crack or become distorted in shape while being burnt, and thus rendered useless. In these cases, it becomes necessary to wholly or partly counteract the contraction.

This is effected by mixing with the clay substances which either do not contract, or which actually expand when heated. To the former class belong burnt clay (vermed grog), coke-dust, graphite, etc., and to the latter class silica, sand, and flint. Ground flint is principally used in pottery. A common mixture for making clay crucibles and retorts consists of two parts by measure of raw fire-clay or a mixture of various clays, and one part of ground crucibles, etc., or other burnt fire-clay.

Analyses of Fire-clays, etc.

	1.	2.	3.	4.	5.	6.
Silica	46·6	46·32	63·3	69·25	98·31	89·04
Alumina	39·5	39·74	23·3	17·9	0·72	5·44
Potash			—	—	0·14 (Potash and Soda)	
Soda			—	—		
Lime		0·36	0·73 (Lime and Magnesia)	1·3 (Lime and Magnesia)	0·22	0·31
Magnesia		0·44				0·17
Ferrous oxide		0·27 (Ferrous and Ferric oxide)	1·8 (Ferrous and Ferric oxide)	2·97 (Ferrous and Ferric oxide)	0·18	
Ferric oxide						2·65
Water, etc.	13·9	12·67	10·3	7·58	0·35	2·3
	100·0	99·8	99·43	99·00	99·92	99·91

1. $Al_2O_32SiO_22H_2O$. 2. China clay. 3. Stourbridge clay (Tookey). 4. Newcastle fire-clay (Richardson). 5. Dinas clay (rock) (Weston). 6. Sheffield ganister.

Fire-clays should be as free as possible from iron pyrites, FeS_2, as this body heated in air yields ferric oxide (Fe_2O_3), which, in contact with reducing agents such as the fuel, is reduced to the lower oxide FeO. This rapidly attacks the clay, forming at the point fusible complex silicates, and the surface becomes pitted, or even covered with a dark-brown slag.

The presence of organic matter is a common occurrence, as these clays are generally the under clay of coal-seams. They are usually hard and rock-like, with a somewhat soapy feel. The bituminous matter colours them grey.

Fire-bricks, besides being refractory, must be strong and of uniform size. The refractoriness is ascertained by making a test-piece from the clay, the edges of which are kept as sharp as possible. This, after careful drying, is strongly heated, and, after cooling, the edges examined. If they remain perfectly sharp, the clay is refractory up to the temperature at which it was heated.

Any softening is evidenced by the rounding off of the edges, and glazing of the surface.

The resistance of fire-clay to fluxes varies with its composition and character. The efficiency with which it is mixed and mellowed prior to use exerts considerable influence on the tenderness or otherwise of the bricks made from it.

Size is an important item in the usefulness of the bricks. If not uniform, much thicker joints will be required in setting them, and as these **joints** are the **weakest part of the lining,** the thinner they can be made the longer the lining will last. Fire-bricks are set in good fire-clay, and not in ordinary lime mortar. The action of lime, if used for this purpose, is obvious. It would combine with the clay at high temperatures, and flux it off—literally run away with the lining.

Ganister.—This substance is a highly siliceous body, as will be seen from the analysis. It is a kind of sandstone in which the grains are cemented by clayey matters, so that, when ground down and moistened with water, it binds together by pressure. Its chief peculiarity is that on burning, it neither expands nor contracts to any great extent. This permits of the lining being formed and burnt in the furnace itself. The moistened material in the form of coarse powder is rammed in between the shell and a wooden core, which is then withdrawn, and the lining gradually heated up. It is used thus for lining the wind furnaces, for melting crucible steel, and the Bessemer converter. It is also used for patching up fire-brick linings. The absence of joints, and the great refractoriness of the body, make these linings very durable. Ganister is also made into bricks. It occurs in the coal measures.

Dinas and Silica Bricks.—The fire-bricks thus known are much more refractory than ordinary fire-bricks, so far as withstanding heat is concerned. As seen from the analysis, they consist mainly of silica. The materials from which they are made differ in character. Dinas bricks are made from a quartzite, and silica bricks from a more granular material of similar composition. The materials are crushed and mixed with 1 to 3 per cent. of milk of lime. This mixture is moulded in iron moulds having a false bottom, with the aid of pressure. After careful drying, they are fired at a very high temperature for several days. During the firing, the small quantity of lime added unites with the silica, etc., at the *surface* of the particles only, and frits or fuses according to the temperature, thus forming a cement, in which the particles of infusible silica are embedded.

NOTE.—The quantity of lime added does not affect the fusibility of the general mass. Its action is restricted to the surface of the particles.

Dinas bricks break with a coarse hackly fracture, in which the milky particles of quartz can be distinguished from the yellow matrix in which they are embedded. Silica bricks have a coarse granular fracture, and feel harsh to the touch.

These bricks are tenderer than fire-bricks, and should be protected from moisture. They *expand* strongly when heated, and hence their application is restricted to those positions where this can be allowed for, or provision made to prevent mishap. Their principal applications are for constructing the ports and roofs, etc., of regenerative furnaces, and roofs of reverberatory furnaces. Consisting as they do of silica, they are unsuited for those parts of a furnace which are in contact with basic and highly corrosive materials or slags. (See Basic Lining.)

Sand is extensively used for making the bottoms of furnaces. The sand employed for this purpose is nearly pure silica. It is used for the bottoms of regenerative open-hearth steel furnaces, and for copper-smelting furnaces. In use it becomes impregnated with metallic oxides, and forms a firm durable lining. Certain sandstones were formerly employed in blocks for the hearths of blast furnaces. The practice is now abandoned, owing to the tendency of blocks of natural stone to crack by heat.

Soap-stone and serpentine are used in Styria and Carinthia for lining the blast furnaces. These substances are hydrated silicates of magnesia, and are highly refractory. They abound in those neighbourhoods. In Sweden that part of the blast furnace which is subject to the strongest heat is lined with a mixture of crushed quartz and clay.

The materials hitherto considered, it will be observed, are of a **siliceous** or **acid** character, and in virtue of their chemical nature are unsuitable for certain purposes, as, for example, where heated for a prolonged period in contact with metallic oxides, which flux them away. Another and more important case, is in making steel from pig iron containing *phosphorus*, in the open-hearth and Bessemer processes. In the purification of iron, phosphorus is removed by oxidation as phosphate of iron —a compound of phosphoric acid and iron oxide—in the slag. This compound is decomposed by silica, which combines with the oxide of iron, forming silicate, and separating the phosphoric acid, which is immediately reduced, and the phosphorus returned into the iron. It thus becomes impossible to remove phosphorus in a furnace lined with siliceous materials. As more than two-thirds of British iron contains too much phosphorus to be used for steel-making in acid-lined furnaces, its removal is a matter of the greatest moment. This can be effected by replacing the acid—siliceous—lining with a **basic** one, *i.e.* a lining consisting of metallic oxides.

Few substances of this nature are available, either from scarcity and consequent high cost, or from lack of refractoriness. They are devoid of binding power.

Among metallic oxides, lime (CaO), magnesia (MgO), alumina (Al_2O_3), and chromic oxide (Cr_2O_3), are most refractory.

Lime, when exposed to the atmosphere, absorbs moisture, forms the hydrate CaH_2O_2, and falls to powder. Its application is, therefore, very limited. It is employed in blocks for the fusion of platinum by the oxy-hydrogen blowpipe.

Magnesia is free from the drawback of absorbing moisture, and the heavy dense form, obtained by strongly calcining magnesite, the natural carbonate of magnesia, forms an excellent lining material, $MgCO_3 = MgO + CO_2$. It is, however, devoid of binding power, and hence something must be employed as a cementing body. Its principal use is for forming the bottom of basic open-hearth furnaces, and for lining basic Bessemer converters. For the former purpose, the strongly calcined magnesite is either (1) ground down to a coarse meal, and then mixed with a small quantity of slag from the furnace, previously ground as fine as flour. This mixture is introduced into the heated furnace in layers, when the slag softens and agglutinates the mass: the quantity of slag is not sufficient to affect the refractoriness of the whole (compare manufacture of Dinas Bricks); or (2) the material may be employed in the same manner as dolomite (see below).

Dolomite.—The amount of magnesite available is small, but, fortunately, the property of not absorbing water applies not only to magnesia, but to the mixture of lime and magnesia obtained by calcining dolomite (mountain limestone). This consists of carbonates of lime and magnesia, and when strongly calcined, the carbonic acid gas is expelled, and a mixture of lime and magnesia remains, which is not readily affected by atmospheric moisture. This substance is largely employed for the purposes stated above, and is commonly known as the **basic lining.** The material is produced in the densest form possible by calcining at about the melting-point of cast iron, with blast and hard coke, so that the maximum shrinkage takes place before its employment in linings. It contracts about 50 per cent., and loses nearly as much in weight. Like magnesite, it has no binding property, and is used by mixing the coarsely ground material with from 10 to 15 per cent. of *well* boiled tar, into a more or less sticky mass, something like asphalte. This mixture, known as "slurry," is rammed into position with heated rammers, round a heated iron core, in Bessemer vessels, and in the bottom and sides of the Siemens's furnace. On heating the lining, the tar is decomposed, or coked, and the carbon remaining cements the whole more or less firmly together. In use, the lining becomes firmer and less porous by impregnation.

Attempts to make the mixture into bricks and burn them are only partially successful, owing to curvature during coking, which prevents them being properly set. Clay, soluble silicate, etc., have also been used as binding agents. Another difficulty is the provision of a suitable material for setting. The introduction of this lining is due to the energy of Messrs. Thomas and Gilchrist, and its most recent application is in copper-refining furnaces, where it promotes the removal of the arsenic from the metal, and diminishes loss, lime and magnesia replacing copper in the slag. The most suitable composition for practical purposes, which shrinks least, is stated by the above workers to be—

Lime	52 per cent.
Magnesia	36 ,,
Silica	8 ,,
Oxide of iron and alumina	4 ,,

Pure **alumina** is known as *corundum* and *emery*, substances whose other properties as gems (ruby, sapphire) or as grinding materials, on account of hardness, enhances their value, and precludes their use as refractory materials.

Bauxite.—A mixture of hydrated alumina and ferric oxide, however, occurs, and is known as *bauxite* (from Beaux in France). Its composition is very variable; the alumina ranges from 35 to 75 per cent., the oxide of iron from 2 to 38 per cent., and the water from 10 to 30 per cent., while silica is present in quantities of from 1 to 15 per cent. On heating, the water is expelled. This material is made into bricks by mixing it, after calcining, with a little clay and some graphite or coke-dust. The clay binds the mass together, and, when burnt, the coke-dust probably partly reduces the F_2O_3 to FeO, which combines with the alumina and forms a highly infusible aluminate of iron, thereby increasing the tenacity of the brick. These bricks have been used successfully in the bottom of basic-steel furnaces, for the lining of Siemens's rotary furnaces, and to form a parting between the basic dolomitic bottom of a furnace and the silica brick sides.

If these are in contact, the lime and magnesia in the lining and the slag, attack, flux off, and undermine the side walls (see Fluxes), with the result that the furnaces collapse. By separating them by a course or two of bauxite bricks this is avoided. Being basic, they are not themselves attacked, and their dense character and composition prevents them attacking the bricks above. Hence the term **neutral course**, which is applied to this parting. Bauxite bricks are also used to line mechanical furnaces for various purposes.

Instead of bauxite, **chromite** is sometimes employed for this purpose. This is a mixture of oxides of iron and chromium, and is very refractory. It is employed in the same manner as dolomite, being either made into bricks or rammed in.

Oxides of Iron.—Besides the above basic materials, various substances, consisting mainly of oxides of iron (Fe_2O_3 and Fe_3O_4), are employed in making the bottom and sides of puddling furnaces for the conversion of cast into wrought iron. They not only serve as a more or less effective protection to the furnace, but play a most important part in the purification of the iron, and will be best studied in connection with the process (see p. 126).

Besides these bodies, others are employed in special cases. In cupelling lead, for instance, **bone-ash** (phosphate of lime) is employed. This body is refractory, and is not readily attacked by oxide of lead. It is also of an absorbent character. In Germany and elsewhere a mixture of **marl** (a clay containing much lime) and charcoal is employed for the same purpose, under the name of *braque.*

Of late years the lining of the blast furnaces used in lead and copper smelting, with siliceous and other materials, has been largely abandoned in favour of water-jacketed furnaces. The highly corrosive slags produced in these processes have little or no action on the water-jacketed iron. Water-cooled iron blocks are also often built into furnace structures, to prevent those parts which are subject to the most intense heat being unduly affected.

Plumbago (natural graphite), being a form of carbon, is quite infusible. Its principal use is in making crucibles, etc. The mineral, after grinding, is treated with hydrochloric acid, to remove the oxide of iron, then washed and mixed with enough clay to bind the material together, and give the necessary strength. Blacklead crucibles contain from 25 to 50 per cent. of graphite.

Crucibles are more or less cup-shaped vessels of refractory material in which substances are melted. This is generally done in wind furnaces, the pots being surrounded by the fire, and when the contents are melted, the crucible is grasped by tongs and lifted bodily from the furnace, and its contents poured out or teemed. These vessels must therefore be

(1) *Refractory*, to withstand the necessary degree of heat.

(2) *Tough while hot*, so as not to break in lifting out.

(3) Must not crack when withdrawn from the fire and exposed to ordinary temperatures, *i.e.* must be capable of resisting sudden and great changes of temperature.

(4) Must not be seriously attacked and corroded by the materials heated in them, or by the ashes of the fuel.

(1) and (2) depend on the materials employed for making the crucible, the second being generally secured by a judicious mixture of various clays, etc. (3) and (4) depend largely on the grain of the crucible; a coarse-grained crucible is less liable to crack than a fine-grained one. This applies also to the heating up of the crucible; with fine-grained pots the greatest care must be taken. On the other hand, coarse-grained pots are more easily attacked by fluxes and fuel ash, so that these two properties do not attain a maximum in the same crucible.

Three distinct varieties of crucible are employed—

Clay, or white pots;
Plumbago, or blacklead crucibles; and
Salamander, or annealed blacklead crucibles.

Clay pots are made from various mixtures of fire-clay, with the addition of grog (ground pots—see Fire-clay), coke-dust, etc., to counteract contraction.

Plumbago pots consist of a mixture of plumbago with sufficient clay to bind it together. They are largely used for the fusion of metals and alloys, being more refractory and less acted on than clay pots. Relatively, with proper use, they last three times as long as clay pots.

Salamander pots do not require the same careful and gradual heating as the other varieties. They consist mainly of graphite in coarse grains, and are coated with a glaze to prevent them from absorbing moisture. These crucibles can be introduced immediately into a hot fire without danger;

the coarseness of the grain, the conductivity of the material, and the absence of moisture prevent cracking. Small ones are specially suitable for blowpipe furnaces working with air or oxygen.

Crucibles are made of various shapes, materials, and finenesses suitable for different operations.

The triangular form is specially suitable in small sizes for melting down metals. The corners are convenient for pouring.

Circular shallow pots, such as the Cornish copper assay crucible, are suitable when roasting as well as when fusions are to be conducted. In copper assays the materials are roasted in the crucible in which they are subsequently fused.

Such pots are also convenient in separating and collecting by fusion substances whose specific gravities are not greatly different, or which do not become perfectly fluid. Pots for tin assay, etc., are of this form.

Deeper pots are preferable where these conditions are not required.

When boiling up of the contents is likely to take place, skittle pots are most suitable. The wide upper portion and contracted mouth prevent the contents from frothing over.

For lifting with basket tongs from the fire, a slight contraction of the top permits of the pots being grasped lower down, and lifted with greater safety.

Fluxing pots are very smooth, and resist the action of such corrosive bodies as oxides of lead, soda, etc., for a considerable time.

Crucible-making.—Small crucibles are made in plaster moulds on a revolving head, or whirling table, somewhat after the manner of pottery. On drying, the clay contracts and loosens itself from the mould, is turned out, thoroughly dried, and afterwards kiln burnt.

Large crucibles are made by hand and machinery. The method of making these pots, for melting steel, at Sheffield is as follows:—A carefully tempered mixture of clays, ground crucibles, and coke-dust is made into lumps of the right size. One of these is placed in a conical iron mould (the flask) previously well oiled. This is provided with a false bottom, having a hole through the centre. A plug—the shape of the interior of the crucible—with a spindle fitting into the hole of the false bottom of the flask, to keep it central, is pressed down into the clay, and by dint of hammering with a mallet, and twisting to and fro, the clay rises and fills the space between it and the flask. When finished, an attendant lifts the whole, and places it on an upright post somewhat smaller than the false bottom. The flask falls by its weight, and the crucible is lifted off and taken away to be dried; or, if the top is to be narrowed, this is done with a sheet-iron cone placed on it and worked to and fro as it stands on the post. The crucibles, after drying, are first carefully heated mouth downward in an annealing oven, some ten or twelve hours being taken to raise them to dull redness. They are then placed, without cooling, on their stands in the fires. These stands are blocks of similar material about 2 inches thick. When fully heated, a handful of sand thrown into the pot, frits, fills up the hole, and cements the pot to the stand.

Large clay crucibles cannot be heated again with safety after being allowed to become cold. They are ground up, after breaking off adherent slag, and used in the manufacture of others, and, in admixture with other materials, for steel-casting sand.

Brasqued Crucibles.—For purposes where contact with siliceous matters is objectionable, crucibles are frequently lined with carbon by mixing lampblack with a mixture of equal parts of treacle and water to a stiff paste. This is rammed into the crucible until it is filled, and a cavity cut out, leaving a lining from $\frac{1}{8}$ to $\frac{1}{2}$ an inch thick, according to size. The crucibles are filled with charcoal, or closely covered and heated to redness. Starch, gum, or oil may be substituted for the treacle, and with large crucibles tar may be employed.

Magnesia or alumina linings may be employed where carbon would be objectionable.

CHAPTER V.

FUELS.

HEAT, for practical purposes, is produced by the combustion or burning of substances in air, or occasionally in pure oxygen. The substance burnt combines chemically with the oxygen, producing gaseous or solid compounds, which pass away to the flues, or, if solid, remain behind. The chemical force exerted in the act of combination appears as heat, and the amount generated is in some measure an indication of the stability of the compound formed.

Any substance which by oxidation is made a source of heat for practical application is classed as a fuel. Most substances, including all those commonly employed, such as wood, charcoal, peat, coal, coke, and gas are derived, directly or indirectly, from vegetable matter, and may be described as **organic fuels.**

Other substances less generally regarded as such are, however, fuels in certain operations. In calcining iron pyrites (which contains 54 per cent. of sulphur) and other rich sulphides, say in a Brückner calciner (p. 29), when the operation is once started, the heat developed by the burning sulphur is sufficient to carry on and complete the calcination.

In this case sulphur is a fuel. In the Bessemer process (p. 147) for making steel, cold air is blown through molten pig iron, and the impurities present oxidized out. Instead of being cooled by the air, the metal becomes very much hotter, chiefly by the oxidation of the silicon in the pig iron to silica ($Si + O_2 = SiO_2$). In the *basic* Bessemer process (p. 152), phosphorus takes the place of the silicon in the ordinary process as a heat producer ($P_2 + O_5 = P_2O_5$). The SiO_2 and the P_2O_5 pass into the slag in combination with metallic oxides as silicates and phosphates respectively. In these cases, silicon and phosphorus are the fuels consumed, and sufficient heat is produced to maintain the purified (wrought) iron in a fluid state.

Sulphur, silicon, and phosphorus may be classed as **inorganic fuels.**

Organic fuels consist mainly of carbon and hydrogen, with varying amounts of oxygen and nitrogen, together with more or less inorganic matter, which is left behind on burning, and which constitutes the ash.

Carbon and hydrogen being the only substances present which burn, a consideration of them is of the greatest importance.

When oxygen occurs in a fuel, it must necessarily be in combination with other constituents. Such part of the fuel as is *already* oxidized cannot be further employed to develop heat, since the heat is produced in the act of oxidation.

In considering the value of a fuel from its chemical composition, it will be therefore necessary to deduct from the carbon or hydrogen a sufficient quantity to combine with the oxygen present. It is usual to make this deduction from the hydrogen. When hydrogen combines with oxygen it forms water, thus—

$$H_2 + O = H_2O$$
$$\textit{parts by weight}\ 2 + 16 = 18$$

or 1 part of hydrogen combines with 8 of oxygen to form 9 of water. Conversely, 8 parts of oxygen require 1 of hydrogen, and by dividing the percentage of oxygen in the fuel by 8, we obtain the amount of hydrogen with which it is combined. Thus, if the fuel contains 18 per cent. of oxygen and 5 per cent. of hydrogen, $\frac{18}{8} = 2{\cdot}25$ parts of hydrogen combined with oxygen *in* the coal, so that only $5 - 2{\cdot}25 = 2{\cdot}75$ parts of hydrogen can be burnt. This is known as *disposable* or *available* hydrogen.

Calorific Power.—When substances unite chemically, the combination always takes place between definite proportionate quantities of the bodies, thus—12 parts by weight of carbon always combine, when completely oxidized, with 32 parts of oxygen, and produce 44 parts of carbonic acid gas, or—

$$\underset{12}{C} + \underset{32}{O_2} = \underset{44}{CO_2}$$

It is equally true that a definite amount of heat is generated. This can be expressed numerically. In burning 12 parts of carbon in the form of purified wood charcoal, 96,960 units of heat[1] are evolved. In burning 2 parts of hydrogen, 68,924 units of heat are given out.

The **quantity of heat** produced in completely burning **1 part by weight of the fuel is the calorific or heating power of the fuel.**

This depends to some extent on the condition of the substance. Compare the calorific powers of carbon as charcoal, diamond, and graphite, in the table given below. The difference is due to the different amounts of heat required to bring about the molecular changes taking place in burning.

TABLE OF CALORIFIC POWERS.[2]

Hydrogen	34,462	Carbon monoxide	2403
Marsh gas (CH_4)	13,063	Sulphur	2261
Charcoal	8080	Olefiant gas (C_2H_4)	11,857
Graphite	7797	Silicon	7830
Diamond	7770	Phosphorus	5747

The calorific powers of the constituents of a fuel being known, it becomes possible to calculate the calorific power of a fuel from its composition.

Example.—A sample of coal gave on analysis, carbon 75 per cent., hydrogen 6 per cent., oxygen 15 per cent., nitrogen and ash, etc., 4 per cent. The available hydrogen = oxygen $- \frac{15}{8} = 6 - 1\cdot875 = 4\cdot12$, and calorific power of fuel $= \frac{75 \times 8080 + 4\cdot12 \times 34462}{100}$.

Calorific powers calculated from analyses of fuels are not reliable, as we have no knowledge of the manner in which the constituents of the fuel are combined.

Direct determinations of the heating power are consequently made. A weighed quantity of the fuel is burnt, and the heat

[1] A unit of heat is the amount required to raise unit weight (say lb.) of water through unit temperature (1°).

[2] The numbers given in this table are the units of water heated through 1° *Centigrade.*

generated is given up to a known weight of water, the temperature of which is previously ascertained. The temperature of the water is again taken after burning the fuel, and the number of degrees it has risen noted.

$$\text{Then } \frac{\text{weight of water} \times \text{rise in temperature}}{\text{weight of fuel}} = \text{C.P.}$$

In the weight of water allowance must be made for the heat absorbed by the vessel containing it, and other parts of the apparatus; and for strict accuracy for other minor losses of heat, such as the heat carried off in the gases as the temperature of the water rises, radiation, etc. For practical purposes these are insignificant if ordinary care is taken.

The instruments employed in making these determinations are known as **calorimeters**, or fuel testers.

Thomson's Calorimeter is shown in Figs. 19, 20, 21. It consists of a glass vessel 12½ inches high and 4 inches wide,

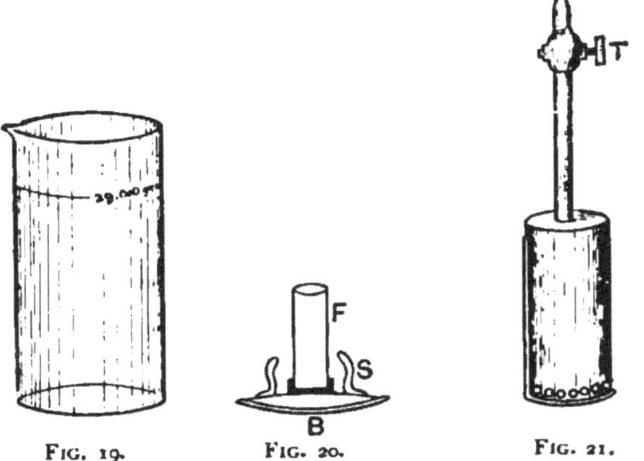

FIG. 19. FIG. 20. FIG. 21.

containing up to the mark 29,010 grains of water. The fuel to be tested, mixed with oxidizing agents (see below), is carefully introduced into the copper furnace tube, F. This is placed in the socket on the base, B, which also carries three springs, S, for the attachment of the cylindrical copper hood (Fig. 21). At the bottom of this hood is a circle of small holes to allow the gases generated to escape, and a narrow tube terminating in a tap, T, rises from the top.

The fuel is ignited by a short piece of lamp-cotton soaked

in nitre, which forms a slow-match. This is embedded half-way in the mixture. As soon as it is ignited, the hood, *with the tap turned off*, is placed over it, and the whole immediately lifted and lowered into the water. The combustion soon extends to the mixture, which burns rapidly. The gases produced bubbling up from the holes, through the water, are cooled. When they cease to come off, the tap is opened, the water rises inside, and cools the copper furnace and body. After gentle stirring, the temperature is taken, and calculation made as before.

The oxidizing mixture employed consists of 3 parts of potassium chlorate and 1 part of potassium nitrate. They must be intimately mixed, quite dry, and in fine powder. The quantity required to burn a sample varies from 7 to 13 times its weight. From 10 to 12 times serves for bituminous coals. In order to obtain oxygen, these substances have to be decomposed, and the escaping gases carry off some heat. The loss due to these causes is estimated at one-tenth of the heat observed. It is necessary, therefore, to add 10 per cent. to the observed rise in temperature. The thermometer employed is exceedingly delicate, reading to $\frac{1}{10}^{\circ}$ F. To ensure steady combustion, the filling of the cylinder must be very carefully done, not too much pressure being employed.

These instruments are designed to take 30 grains of the fuel, and were primarily intended for testing fuels for boiler purposes.

The number 29,010 was selected with a view to make the rise in temperature indicate the evaporative power of the fuel. The latent heat of steam is 967° F. units, and $967 \times 30 = 29{,}010$. Thus, if 30 grains of fuel heat 29,010 grains of water 1°, 1 grain heats 967 grains of water 1°, and would therefore convert 1 grain of water at 212° F. into steam at 212°. Hence the rise in T° + 10 per cent. = evaporative power.

The calorific power in heat units is obtained as before.[1]

NOTE.—The calorific powers given on p. 44 were made by Faure and Silberman, with a most delicate calorimeter, for which see Ganot's "Physics," pp. 401, 423.

In burning hydrogen, the product of combustion—water—

[1] If a Fahrenheit thermometer is used, the results can be converted into Centigrade units by multiplying the rise in the temperature by $\frac{5}{9}$, and *vice versâ*.

is a liquid at ordinary temperatures. The calorific power given above includes the total amount of heat given out. At the temperature of burning, the water is in the state of steam. To retain it in this condition,[1] 637 units of heat per part of water will be rendered latent and unavailable for heating purposes. Since each part of hydrogen produces 9 of water, $637 \times 9 = 5733$ units must be deducted for this purpose. The remainder, 28,729, is the heat available for raising temperature. This deduction applies to all moisture present in the fuel as well as that formed in burning.

In burning carbon, note should be taken that carbon forms two oxides, CO and CO_2. The calorific power of carbon burning to CO is only 2473, less than a third of its total heating power.[2] This shows the importance of complete combustion in order to secure economy.

The temperature produced by burning a given fuel is dependent not only on the amount of heat given out, but on other conditions also: the *amount* and *nature* of the products of combustion, whether the combustion takes place in air or pure oxygen, and the initial temperature. The degree of heat realized is below that calculated, and is determined by the stability of the products of combustion, for, when a certain degree of heat has been obtained, these are dissociated as rapidly as formed, and the heat thus absorbed is balanced by that given out.

In practice, the temperature, when burning solid fuel, depends on the rapidity of combustion and the density of the substance, assuming that the composition is the same. The more rapid combustion which attends the employment of hot air, as well as the heat carried in by the air, greatly increases the temperature, while the structure of the substance influences the rate at which it burns—porous cellular bodies burning most freely.

Dense fuels, when burning at the same rate as lighter ones,

[1] This is made up of the latent heat of steam (537) and the heat required to raise water to boiling-point (100 Centigrade units).

[2] The remainder of the heat is given out when the CO burns to CO_2 $CO + O = CO_2$.

produce greater local heat, the heat-evolving and radiating power being concentrated in smaller volume.

Wood is extensively employed as a fuel where a plentiful supply is obtainable and high temperatures are not required.

The organic constituents of dry wood, exclusive of ash, are—

Carbon	51 per cent.
Hydrogen	6 ,,
Oxygen	41·5 ,,
Nitrogen, etc.	1·5 ,,
	100·0

NOTE.—The composition of various kinds of wood is almost identical, no constituent varying much more than 1 per cent. The principal body present in all is cellulose $C_{12}H_{20}O_{10}$, with various hydrocarbon substances, as turpentine, resins, etc., which influence its inflammability. Its density varies from 0·4 to 1·3. The large amount of oxygen will be at once noticed, and from what we have previously learnt, the disposable hydrogen is only $6 - \frac{41 \cdot 5}{8} = 0 \cdot 82$ per cent. All the water of composition, $41 \cdot 5 + 5 \cdot 18 = 46 \cdot 68$ per cent., has to be evaporated. This, added to the fact that ordinary air-dried wood retains from 15 to 20 per cent. of mechanically held moisture, which has also to be expelled, will show the unsuitability of wood for the production of high temperatures. If kiln-dried, it reabsorbs a great part of the moisture on exposure.

The ash of wood rarely exceeds 2 per cent. It is characterized by the presence of a considerable amount of potash and the absence of alumina. It consists of carbonate of potash, lime, soda, iron, magnesia, with a little chlorine, sulphuric and phosphoric acids, and silica. Wood ashes formerly formed the chief source of potash salts.

The woods principally employed are: larch, fir, sycamore, birch, elm, ash, pine, and oak.

The inflaming point of wood is about 300° C., much below redness.

Charcoal.—When wood is gradually heated out of contact with air, it undergoes a destructive distillation. Water and various other volatile compounds are expelled, some of which result from decomposition of the cellulose and other bodies present in wood, with the separation of free carbon. This decomposition commences at about 180° C., and is completed at about 400° C. The residue obtained is charcoal. It consists of the *fixed* carbon of the wood, with the ash, and some hydrogen and oxygen, the amount of which depends on the temperature of preparation.

The substances expelled consist mainly of water, wood naphtha, various heavy hydrocarbons constituting tarry matters, marsh gas, hydrogen, olefiant gas, carbonic oxide, carbonic acid gas, pyroligneous acid (crude acetic acid), and ammoniacal compounds. The valuable nature of some, and combustible character of others of these substances, will be noted.

The weight of charcoal obtained varies from 15 to 25 per cent., rarely exceeding 20 per cent. Its volume is from 50 to 75 per cent. of the wood. The yield depends on the nature of the wood, the heat employed, and rapidity of charring. High temperature and slow charring diminish the yield owing to the more complete distillation which occurs. Good charcoal should be hard, sonorous, give a bright fracture, not soil the hands, not friable or fissured, and retain the form of the original wood. Its igniting-point depends much on the temperature of preparation, as the higher the temperature the denser and less easily ignited it becomes. Rapid charring has the effect of making the charcoal fissured.

The combustible nature of the substances expelled by charring will show that, unless high local heat is required, it would be more economical to simply dry the wood. The *quantity* of heat given out by burning the wood itself is greater than that given out on burning the charcoal prepared from it.

Charcoal may be prepared in two ways. The wood may be charred in retorts heated externally by a fire, or it may be piled in a kiln or stack, and the charring effected by a complete or partial burning of the volatile matters which distil off when the wood gets heated. The preliminary heating is effected by faggots put in some convenient position in the pile.

Charring in retorts is principally followed for the sake of the pyroligneous acid and tar, the charcoal being a by-product.

Charcoal Burning in Piles.—The piles are made either circular or rectangular. In circular piles, the wood, sawn into suitable lengths, is built up round a central stake or stakes, in the manner shown in Figs. 22, 23, and the pile covered with sods or earth, supported by branches, whose ends are stuck in the ground and bent over, or, sometimes by a mixture of charcoal and water. This furnishes a yielding cover, but one sufficiently impervious to air as to exclude excess. If three central stakes are employed, the chimney thus formed is filled with faggots. If only one stake is used, a passage is left at one side reaching to the middle, and this is similarly filled. The upper part of the heap is made up with branches and irregular pieces as solidly as possible. When the pile is completed, the faggots are ignited and the openings left uncovered until the pile has fairly caught. They are then completely closed, and the pile left to itself. Volumes of dense yellow

FIG. 22. FIG. 23.

smoke are at first given off, accompanied by much water vapour. This condenses in the cover and runs down. After a time, the yellow smoke changes to grey, and the cover is extended down to the ground, leaving only a few small openings judiciously arranged to admit a little air, to continue the combustion of the volatiles, and maintain the heat. The now thoroughly dried wood is thus gradually converted into charcoal, and the "coalier," or burner, completes the charring of the outer portions of the heap, or any part that has not caught well, by making a series of openings in the cover, commencing near the top, which draw the fire and heat in that direction. The smoke which first escapes gradually becomes thin, and the flame of CO appears. When this occurs, the holes are stopped up, or the charcoal will burn, and a new series opened lower down. This is repeated till the whole pile has been charred.

The judicious arrangement of the wood so that the combustion shall spread uniformly, the consolidation of the pile and making good of any falling in due to contraction in the earlier stages, and the proper management of the vents, are necessary to secure a satisfactory result.

The heat is maintained by the combustion of the volatiles inside the pile. If excess of air is admitted, the charcoal will be partly burnt. The quality is said to be improved by quenching it before it has cooled below its igniting-point. This is accomplished by taking out some of the charcoal, through an opening made in the cover, which is at once replaced. The portion removed is cooled by water or by wet sand, earth, or charcoal-powder. The quenching prevents burning of the charcoal, which might occur during the cooling, if the stack were not tight.

In rectangular piles (Fig. 24) the wood stack is first built, and surrounded with planks supported behind by stakes driven into the ground, a space being left between the pile and the inner side of the planks. This is filled with charcoal-dust

Fig. 24.

moistened with water, or ashes, to protect the planks from the heat. The top is covered with earth, ashes, etc., and ignition is effected through the opening shown in the lower end. The charring proceeds as before. These piles are 22 feet long, 4 broad, and from 7 to 9 feet high. Much of the acid and tarry matters can be collected by iron pipes introduced through the cover at the higher end of the pile, leading to receivers. These piles are generally built upon sloping ground. They are more common in Norway and Sweden than elsewhere.

Where a continuous supply of wood is obtainable (as, for example, on a lake shore or river-side, by floating the wood), a permanent masonry bottom is built for the piles. This slopes to a cavity in the centre loosely covered with an iron plate, and communicating by a passage with a tar well. The condensed tar and pyroligneous acid drain down and pass into the well.

Rectangular kilns of masonry are sometimes employed, provision being made for igniting the wood, for leading away the products of distillation, and for regulating the air supply.

Wood for charcoal burning should be mature, but not decayed or worm-eaten. It is at its best when about thirty years old. It should be felled in winter, when the sap is down, and the bark removed to facilitate drying. The site selected should be near a stream or water supply, and the ground not too sandy or clayey. The former is too porous, and the latter cracks with the heat, and air is drawn in. If burnt, charcoal is dull, soils the hands, and is light and friable.

The amount of charcoal produced varies with the method of burning. It forms from 14 to 25 per cent. by weight, and from 50 to 75 per cent. by volume, of the wood. It absorbs about 10 per cent. of moisture on exposure to the air.

The specific gravity of charcoal varies from 0·11 to 0·2. If, however, air is expelled from its pores, it is found to be 2.

The **Composition of Charcoal** varies with the temperature of preparation. Ordinary charcoal, prepared between 400° and 1000° C., is as follows:—80 to 83 per cent. carbon; 1 to 2 per cent. hydrogen; 14 to 15·5 per cent. oxygen and nitrogen; 1 to 5 per cent. ash.

Peat, or Turf, is produced by the gradual accumulation of dead vegetable matter, especially mosses and lower forms of vegetable life, in moist situations. The moisture protects it from the action of the air. Under these circumstances, a gradual change in its composition goes on. The oxygen and hydrogen in the original vegetable matters are slowly eliminated, as water (H_2O), marsh gas (CH_4), carbonic acid gas, etc. Up to a certain point the oxygen is removed in greatest proportion, the hydrogen in a less degree, and the carbon in the least proportion. The net result of these changes is that the proportion of carbon increases, the colour darkens, greater density is attained, and up to a certain limit the disposable hydrogen is increased.

These changes invariably occur when vegetable matter undergoes alteration in absence of air, and are assisted by even very moderate degrees of heat, such as the internal heat of the earth. The marsh gas, causing on ignition the will-o'-the-wisp, is produced in this manner. The fire-damp of the mine is also marsh gas, which has been stored in the coal under the pressure of the superincumbent material, and which, on opening up the seam, diffuses out of the coal. CO_2 is more rarely met with owing to its solubility in water, which is always filtering through the rocks more or less.

The greater the degree of alteration the more widely it departs in character from the original vegetable matter. It is in virtue of these changes that coal, some of the varieties of which consist almost entirely of carbon, have been produced from the vegetable deposits of former ages. Under similar conditions, the greater the age the more altered the material becomes.

Peat usually occurs at the surface, filling up basin-like depressions. These are known as bogs, or mosses. It follows that *peat*, being of recent origin, is comparatively little changed, and that the upper and newest layer will differ from the lower and older layers. It will consequently resemble wood in chemical composition. On drying, peat from the upper part of a bog yields a light brownish-yellow fibrous substance, forming about 70 per cent. of the volume, and retaining sometimes over 30 per cent. of moisture after air-drying. Peat from the bottom of the bog is more gummy, and on drying yields a dense black compact mass, forming about 27 to 30 per cent. of the volume, and retaining about 20 to 30 per cent. of moisture after air-drying. The specific gravity of peat varies from about 0·1 to nearly 1; as taken from the bog it contains from 70 to 90 per cent. of moisture. The peat after removal is dried on floors, built in walls, and afterwards stacked or housed. This should be done in open weather, as frost seriously injures fresh peat. It never drys so firmly and dense after being frozen.

The **Preparation of Peat,** so as to produce from it a denser material more suitable for use as fuel, has received much attention.

Most of the methods adopted involve compression of the peat, either in a wet or air-dried state, into blocks. In others, the peat is converted into pulp by grinding. This, on drying, contracts to about one-fifth of its bulk, and yields a much denser material containing less moisture.

The **Ash of Peat,** as would be expected from the situations in which it is found, is much higher than that of wood. It ranges from 8 to nearly 30 per cent. It contains the same constituents as wood ash, with the addition of alumina. It contains, also, more of sulphates and phosphates, and often sulphides as well.

Peat commences to distil at about 130° C., and leaves a charcoal, the value of which depends on the character and amount of ash of the peat.

Fossil Fuels.—When deposits of vegetable matter have, by alteration of the earth's surface, been submerged in the sea, and other strata deposited on them, the alteration previously noticed has continued, and the substance has entirely lost its vegetable character and become fossilized. The extent to which this has gone on depends on the age of the geological formation in which it is found, and sometimes on local influences.

Those substances which are found in the newer formations are called *lignites* (Lat., *Lignum* = "wood"), from the distinct woody character of some of them, and those found in the older formation—known as the "Carboniferous"—are called *coal.* As would be expected, much difficulty is experienced in drawing a sharp limit between the two, as they gradually merge into each other.

Characteristic specimens, however, differ widely. Selected examples of lignites and coals show a gradual passage from wood to anthracite (the most altered form of coal). Such a table is exhibited below. The striking feature of such an arrangement is that the available hydrogen gradually increases to a certain point, and the amount of fixed carbon—carbon not driven off when the coal is heated—shows a similar increase.

The effect of this upon the character of the coal is important. The higher members of the series, in which the available hydrogen is low, burn without softening and fritting together, and if the powdered fuel is heated in a vessel from which air is excluded, the particles do not stick together. Such substances are described as **non-caking**. As the available hydrogen increases, the caking property becomes more and more strongly marked, until the substance becomes so rich in carbon as to again become non-caking, the bituminous matters produced during heating not being sufficient to bind the particles together. Thus the non-caking substances are of two classes: (1) those rich in oxygen, and low in available hydrogen; (2) those rich in carbon.

COMPOSITION OF FUELS.

Fuel.	C.	H.	O.	N.	Ash.	Available hydrogen.	
Wood (dessicated)	51·1	6·2	41·4	1·12	1·8	1·1	Non-caking.
Peat (Coppage, Ireland)[1]	52·38	7·03	40·59[2]			2·1	
Peat, Long (France)[1]	60·9	6·22	32·88[2]			2·3	
Lignite—							
Caroline, S.	60·3	4·8	20·2	1·0	3·2	2·3	
Auckland[2]	64·7	4·81	18·25	1·34	10·48	2·53	
Tasmania	69·14	5·4	18·48	1·26	5·37	3·1	
Trinidad	75·63	5·2	13·51		2·64	3·5	
Coal—							
Cannel, Wigan	80·07	5·53	8·1	2·1	2·7	4·5	Caking.
Andrew's House, Eanfield	85·58	5·37	4·39	1·26	2·14	4·8	
Blaina	83·0	6·19	4·58	1·49	4·0	5·6	
Ebbw Vale	89·78	5·15	0·39	2·16	1·5	5·1	
Aberaman	90·94	4·28	0·94 O. and N.	1·21	14·5	4·1	Non-caking.
Anthracite (Isère)	94·0	1·49	3·58[2]		4·0	1·1	

Lignites.—Some of these are but little altered from woody matter. They are of a light colour and fibrous structure. These may be classified as *fossil wood*, or fibrous lignite. Such a deposit occurs at Bovey Tracey, in Devonshire. These lignites contain 30 to 50 per cent. of moisture, as won from the ground, and retain 12 to 20 per cent. after air-drying. On heating, they leave a residue of about 35 per cent.

Bituminous, or earthy, lignite is a convenient designation for those more altered in character. The colour is dark brown, the fibrous structure indistinct, and the fracture earthy. They contain less moisture than fossil wood.

[1] Exclusive of ash. [2] Inclusive of N.

The residue left on heating varies from 35 to 50 per cent., and 4 to 5 per cent. of tarry matters. Specific gravity 1·1 to 1·2.

The most altered lignites resemble coal in appearance, and to some extent in character. Some are black and shiny, others are dull, and black-brown in colour. The fracture is flat or conchoidal. All trace of woody fibre is lost. They contain less moisture, and leave a fixed residue up to 60 per cent. on distillation.

This class includes all the better varieties of **brown coal** (German, *Braunkohle*).

The volatile matters expelled from lignite resemble more or less those from coal, but are characterized by a large amount of aqueous distillate. The tars average some 4 to 6 per cent. Lignites are largely used in Germany, France, Italy, and Austria.

The ash of lignite consists mainly of oxide of iron, alumina, silica, and sulphate of lime and iron. It varies from 1 to 50 per cent.

The average composition of the organic constituents of the three varieties given below is from Regnault.[1]

	Carbon.	Hydrogen.	Oxygen and Nitrogen.
Fibrous lignite	63	5	32
Earthy	72	5	23
Pitch-brown coal	77	7·5	15·5

Coal.—Under this heading are included the more altered forms of fossil fuel. The term "bituminous coal" is applied to all those which burn with a more or less considerable amount of flame of a smoky nature, somewhat resembling pitch and bitumen.

Bituminous coal passes into anthracite, which burns without flame, smoke, or smell. Coals which burn rapidly, without softening and fusing together so as to arrest the draught, are called **free burning.**

Caking Coals include all those which soften and stick together when heated. If the powder is heated in a closed vessel, a more or less coherent mass of coke is obtained. *Free-burning coals* are "non-caking," or only slightly so.

[1] See Mills and Rowan's fuel.

As almost every coal-seam differs more or less from others, it is necessary to establish some method of classification.

Since the chemical composition of the coal affords little clue to its behaviour in burning, the most convenient classification is based on the amount and nature of the residue left, when the coal is heated in a closed vessel, minus the ash.

Substances used exclusively for the manufacture of gas (boghead coal, etc.), or oils (paraffin coal), are not included.

Class 1.—Non-caking coals rich in oxygen (Percy). This includes the various kinds of *cannel*,[1] splint, or hard coal. They burn freely, with a long flame like a candle. Cannels possess a dull pitchy lustre, break with a conchoidal fracture, give a brown streak, and are hard and dense. The specific gravity is about 1·2. On heating, these coals retain their form, but do not cake together. The lumps of residue are cracked and friable. The percentage of coke varies from about 40 to 60 per cent. The fixed carbon present in the coke varying up to 53 per cent. Cannels yield on distillation a larger percentage of volatile matters, and less coke than other bituminous coals. The ash and sulphur are also higher. Splint or hard coal is employed in blast furnaces in Scotland and Staffordshire.

The calorific power of these coals, *free from water and ash*, varies from 8000 to 8500. They occur in Staffordshire, Derbyshire, Lancashire, and Scotland.

Class 2.—Caking coal, burning with long flame—cherry coal (fat coals—Gruner). This class includes the various gas and many *steam* coals. These coals ignite easily, and burn freely with much flame and more smoke than cannel. They are very black and bright, and somewhat platey in structure, much more friable and not so hard as cannel, and are known as *soft coal.* Heated in a closed vessel, they coke slightly, some to a greater extent than others. The coke is light, spongy, and friable; it forms from 60 to 70 per cent. of the coal coked.

The gas is of good quality. They are largely used for manufacture of gas, and for steam-raising. The calorific

[1] Cannels are supposed to have been produced in a different manner to coal. Some are caking in character.

power varies from 8500 to 8800. They occur extensively in South Wales and in the Newcastle, Staffordshire, and Glasgow coal-fields.

Class 3.—Caking or soldering coal—smithy coal. Coals of this class almost fuse when heated, and form a pasty mass from which bubbles of gas escape, leaving a coke altogether differing in form from the original. The flame is bright and luminous. The coals have a velvety black colour, generally soil the fingers, and have a tendency to break up into small rectangular pieces. They swell considerably during coking, and this reduces the density of the coke, which varies in quantity from 68 to 74 per cent. The calorific power is 8500 to 9300, but they are unsuitable for steam-raising, and many other purposes, owing to the great tendency to cake and impede the draught. Many Continental coals are of this character. In Britain they occur in Durham, Yorkshire, Lancashire, Staffordshire, Derbyshire, South Wales, and other localities.

Class 4.—Coking coal (fat coal, burning with a short flame —Gruner). This includes those coals which, on account of the large yield and dense nature of the coke, are most suitable for making coke for use in blast furnaces. They are generally of a soft nature, liable to fall to pieces and crush. They ignite and burn less readily than preceding varieties, but do not soften and swell to the same extent. The flame is short, white, and almost smokeless. The coke is denser and stronger than from Class 3, and the yield from 74 to 82 per cent. It is the best for blast-furnace work. The calorific power is from 9300 to 9500. They are less suitable for steam-raising, unless forced draught is employed, than a freer-burning coal. They occur in South Wales, at St. Etienne, and elsewhere.

Class 5.—Non-caking coals rich in carbon (Percy), anthracitic coal. These coals gradually pass into true anthracite. They are harder, burn with little or no flame, are smokeless, and odourless. They ignite and burn with great difficulty, and, unless forced draught is employed, perfect combustion cannot be ensured. Unless slowly heated, they crackle, fly to pieces, and choke the air-ways. Some kinds are less liable to do this than others, and in South Wales and Pennsylvania certain varieties are used for blast-furnace work. In appearance they are dull or streaky, and break with a more or less conchoidal fracture. The calorific power is somewhat less than class 4, as the hydrogen present is less. They leave an uncoked residue of about 82 to 88 per cent. They are employed as steam coals.

Anthracite is the most altered form of coal. It has a brilliant black or semi-metallic appearance, and gives a black streak. It is non-caking, and most difficult to burn, requiring a huge draught. It is the densest of the coals, and generates an intense local heat. It is more liable to fly to pieces on heating than Class 5, bituminous coals. It burns without flame or smell. South Wales, Pennsylvania, and the Vosges are the principal localities. Anthracite leaves from 85 to 94 per cent. of fixed carbon, with less than 5 per cent. of ash.

The **specific gravity of coal** varies from 1·25 to 1·31. It is, however, influenced by the amount of earthy matters present. The **ash** ranges from 2 to 18 per cent., and consists of lime, alumina, oxide of iron, magnesia, alkalies, phosphoric, sulphuric, and hydrochloric acids, and silica.

The **sulphur** present in coal is of the greatest importance in the use of the material for iron manufacture in blast furnaces, as it is taken up by the metal. It exists in three states: (1) as iron pyrites—the brassy material in coal; (2) as organic sulphur; (4) as sulphate of lime, and sometimes of alumina. The two former states are most objectionable, *when used as coal* in iron smelting, as the sulphur, sulphuretted hydrogen, carbon disulphide, and sulphide of iron, which result on heating, may all transfer their sulphur to the metal. By previously converting the coal into coke, the organic sulphur is removed, and about half the sulphur in the pyrites.

To purify the coal from pyrites and dirt, for coking, forge, and other purposes, coal screenings are washed. The inorganic matter, in virtue of its greater density, separating from the coal. The specific gravity of pyrites is 5, nearly four times that of coal.

The pyrites present in coal often contains arsenic, and sometimes copper.

Phosphorus is generally only present in very small quantities. **Chlorine** is always present. It should not be overlooked in coals used for steam-raising in boilers fitted with copper tubes, which it rapidly corrodes.

The selection of a coal for any particular purpose depends as much on its physical as its chemical character. For blast-furnace work it must be hard and strong, not crumbling under the pressure of the charge, and it must not be too strongly caking. It must, moreover, be practically free from pyrites. Certain varieties of Classes 1, 2, and 5 of bituminous coal, and anthracite, are used for this purpose.

In reverberatory furnaces working with draught, and for steam-raising, free-burning coal is employed.

For **technical** purposes, the ash, fixed carbon, volatile matters, sulphur, and calorific power are usually determined.

Coke.—From the above consideration of coal, it will be seen that certain classes are not particularly suited for use in that form, either from their coking power, softness, or the presence of sulphur. These defects may be overcome by converting the coal into coke. Very soft coals often yield excellent coke, and, as before pointed out, half the sulphur in the pyrites present is expelled during the coking, together with the organic sulphur, mainly as H_2S and CS_2. So that many coals unsuitable for iron manufacture yield coke which is not so prejudiced.

Coke stands to coal in the same relation as does charcoal to wood, and consists of the fixed carbon together with the inorganic constituents of the fuel. The percentage of ash in coke is consequently higher than in the coal itself.

When heated out of contact with air, coal splits up, yielding hydrogen, various volatile compounds of hydrogen and carbon, and of these elements with oxygen, ammonia, water, and coke. The heavier of the hydrocarbons, etc., constitute *tar*, and the water and ammonia, *ammoniacal liquor*. The lighter of the hydrocarbons are non-condensible, and constitute coal gas. From tar, bisulphide of carbon, benzol, toluol, naphtha, creosote, phenol, anthracene, naphthalene, and pitch are obtained by distillation, at a gradually increasing temperature.

These substances are very valuable products. The composition of the tar, and the proportions of its constituents, will vary greatly with the temperature of coking. Low temperature favours the production of a tar low in benzol, toluol, carbolic acid, etc., and containing much heavy oily paraffin. A high temperature favours the production of tars rich in benzol, etc. These are much the most valuable. At high temperatures—above 1200° C.—heavy hydrocarbons are decomposed, depositing part of their carbon, and being resolved into lighter bodies and hydrogen. This is often seen in gas-retorts, the inner surface of the retort, with which the gas has come into contact, being covered with a layer of dense carbon thus deposited. This is sometimes graphitic.

If the coal while coking can be sufficiently heated to decompose these in its mass, the coke obtained will be denser, stronger, and more brilliant in appearance. The yield of coke will also be greater, in proportion to the amount of carbon thus retained. It follows that the quality of the coke will depend not only on the nature of the coal, but also on the temperature of coking, and the rapidity with which it is attained, those processes producing the best coke in which the highest temperature is most rapidly obtained.

As in charcoal-burning, the heat necessary may be obtained by partially or completely burning the volatile matters, in contact with the coal, or outside the coking chamber.

NOTE.—In all cases where the burning goes on in contact with the substance carbonized, the air supplied should pass from the unburnt to the burning body.

Coking in Heaps, "Meiler," or Mounds.—The coal is piled up round a temporary chimney loosely built of bricks, with an iron cover-plate to regulate the air supply. In some cases the operation is conducted like charcoal-burning, the heap being covered with earth and moistened coal- or coke-dust, and vents made as required.

In another method no cover is put on till the coking is completed, the pile being ignited on the top and the fire proceeding downwards and throughout the mass, air finding free ingress. The combustible volatile matters distilling off from below, ascend and protect the coke from burning. When a thin film of ash appears on the surface, showing that the coke is burning, a cover of earth, coal- or coke-dust is applied at that part, and thus repeated till the whole is covered and the coking complete. *Coking in long ridges* is a similar process.

Coking in Kilns.—These consist of two parallel walls about 5 feet high and 8 feet apart, and 40 feet long. The two ends are left partly open for charging, and bricked up while coking is going on. The walls are pierced at a height of 2 feet from the ground with openings, from each of which a vertical flue rises to the top. These openings are about 2 feet apart. The kiln is charged by building up one end, wheeling in damp slack, and ramming it down till level with the flues. Passages are then built across the kiln by leaning lumps of coal together, so as to leave a channel or flue, or billets of wood are placed across, and, after charging, withdrawn. The kiln is then filled up, the larger stuff being placed over the air-ways, and the smallest on top. The end

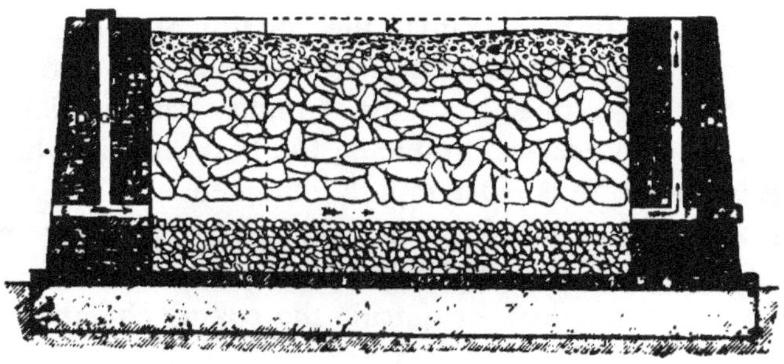

Fig. 25.

is bricked up, and the top covered up with loam and coal- or coke-dust. The coking is started by combustibles (chips, etc.) pushed in through the horizontal flues on one side. The *vertical* flues on *that* side are closed by tiles, and the *horizontal* flues on the *opposite* side (see Fig. 25). On igniting the sticks, the draught passes in the direction of the arrows, and the combustion extends across the kiln. The amount of air supplied is regulated by the opening E. After a time, the vertical flues on the opposite side are closed, and the horizontal flues opened, those open before being stopped. The direction of the air current is now reversed, and by repeating this at intervals of about two hours, the whole mass is gradually coked.

These kilns should be sheltered from wind, which will otherwise interfere with their regular working. The same objection of partially burning the coke applies to these kilns as well as heaps. The operation lasts about eight days. When complete, the end walls are taken down, after cooling, and the coke discharged. It is needless to say that the weather—wet or dry—influences the temperature, and thus the quality, of the coke. The effect of moisture will be seen later on.

Coke Ovens.—Coke is now generally made in ovens, *i.e.* closed chambers.

They may be divided into—

(*a*) Simple chambers, to the interior of which air is admitted, to burn the products of distillation.

1. Cold-air—Beehive, Rectangular.
2. Hot-air (Jones and Cox's).

(*b*) Ovens in which the distillation products are *all* burnt outside the chamber. Appolt and Coppée ovens.

(*c*) Ovens in which the tarry matters and ammonia are removed by condensation from the gases, and the *non-condensible* gases burnt outside the chamber in flues.

1. Those in which some air is admitted to the chamber (Jamieson).
2 Those from which air is rigidly excluded (Pauwell's, Pernolet, Simon-Carves).
3. Those in which the gases are burnt by heated air (Simon-Carves's improved, Bauer, Otto-Hoffmann).

Beehive Ovens.—These ovens are very largely in use. They yield coke of good quality, and can be employed with all classes of coal, whether it swells on coking or not, a consideration not to be overlooked. On the other hand, the yield is less than ovens to which air is not admitted, owing to burning of the coke, and much heat is lost and time wasted. The prime cost is, however, low, it costs little for repairs, and requires no large amount of skill. The chamber (Fig. 26) is circular, 10 to 12 feet in diameter, 2 feet to spring of dome, and 7 feet to the crown from the floor. The ovens are lined with refractory bricks, and are built in blocks of 40 or 50 in a double row, back to back, on a raised platform some 2 feet above ground-level. The block is surrounded by a strong wall, and all spaces are filled up with sand or granulated slag to retain heat. A rail-track runs along the edge of the platform.

Each chamber is provided with a short chimney, or communicates by a short flue with a wide common flue, which runs between the two rows forming the block, and terminates in a stack.[1] The short flues can be closed by dampers, as shown. In front is an arched opening, some 3 feet high, which serves as a door for discharging the coke.

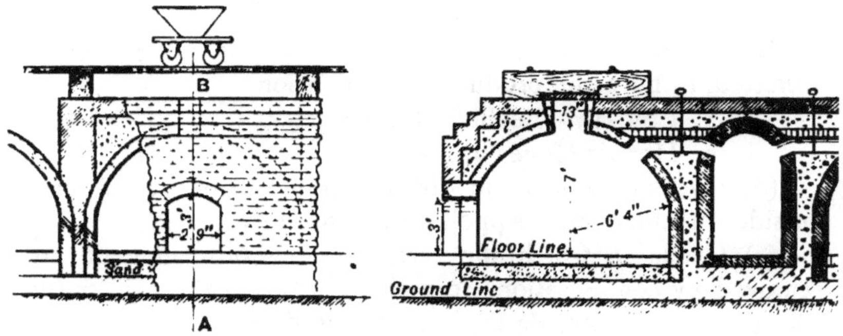

Elevation part in Section. *Section through A.B.*

Fig. 26.

The coal is introduced through the top, from hopper waggons running on a rail track, and raked level.

The charging opening is covered, and the cover luted. In some cases the coal is shovelled in from the front.

The chambers are hot from previous charges. After charging, the front is loosely bricked up. If the oven is hot enough, the bricks are smeared over with loam to exclude air. If not, they are left uncovered for a time, and sometimes openings are made near the bottom to admit air. Distillation commences at once, but the gases are incombustible at the temperature of the oven. In from 1½ to 3 hours ignition

[1] This flue gets intensely heated, and ensures combustion of the products of distillation. The hot gases are passed under boilers, to raise steam, before passing to the stack.

commences, and the gases burn with a long, red, lurid flame, and much smoke. A small opening is then made in the top of the doorway to admit air *above the coal*, and burn the gases in the oven.

The temperature rapidly rises. The dome-shaped roof reflects the heat on to the mass of coal below, and this gradually gets heated through. The gases distilling off from below are partly decomposed in passing through the heated upper layers, and deposit carbon. The air-supply to the oven is regulated so as to burn the products *in* the oven as completely as possible without admitting any excess. When the distillation begins to slacken, the holes in front are stopped one by one, until the doorway is again completely closed. The chimney is also stopped, and the coke is left to itself some 12 hours to complete and cool. The doorway is first taken partly down, a hose-pipe introduced, and the coke quenched with water, in the oven, below its igniting-point. The door is then completely removed, and the coke discharged with rakes and forks. It breaks into columnar masses, the axes of which are vertical. This is owing to the direction of the coking, which takes place downwards. These ovens make from 3 to 5 tons per charge, and the yield is about 60 per cent.

The **rectangular oven** is exactly similar in principle, but the chambers are rectangular. The coking is conducted in the same manner. Sometimes the whole front of these ovens is open, and the bottom is made slightly sloping. It is then possible to remove the coke in one mass. For this purpose, before introducing the coal, a couple of strong iron drag-bars, turned up at each end, are laid on the floor of the oven, with the ends projecting, and the coal charged in on these. When the operation is complete, a windlass is attached to the projecting ends, and the whole mass dragged from the oven, being quenched as it comes forward on to the platform in front. The oven is thus left much hotter, and less heat and time are wasted.

In others, the front is arranged like a beehive, and sometimes an iron frame filled with fire-brick blocks, sliding in guides, and counterpoised, is used to close the mouth of the oven. Two charges a week can be worked off from each chamber. The coking itself occupies about 48 to 60 hours.

Cox's coking oven has a double-arched roof. The gases escape through an opening in the front of the lower arch, and pass back to the flue. No air is admitted in front, but air drawn from the front, through flues in the brickwork, is admitted by openings at the back, above the coal. The air-supply is regulated by a little sliding door over the opening of the flue in front. In passing through the flues it gets heated, and thus a higher temperature in the oven, and more rapid coking result. Some 12 hours are saved by this arrangement.

In **Jones's oven**, the gases, after leaving the chamber, are caused to pass through flues under the chamber, before passing to the chimney, thus heating the coal from below.

All these structures are built in a massive manner, and spaces filled up with sand, etc., to retain as much heat as possible.

In ovens of the second class, air is not admitted to the coking chamber, but the gases, etc., are burnt *outside*, in flues or spaces surrounding the chamber.

In the **Appolt coke oven**, the chambers are tapering, vertical brickwork retorts of rectangular form, 13 feet high, 4 feet by 1 foot 6 inches at base, and 3 feet 8 inches by 13 inches at top. These retorts are built in two rows, 18 or 24 in a block, with a surrounding space varying from 7 to 11 inches wide, and tied together and to the surrounding wall with bricks for mutual support. They are a single brick thick, and are supported on two parallel arches. The retort bottom consists of a hinged iron plate, which can be lowered so as to discharge the coke into the arched vault below. It is

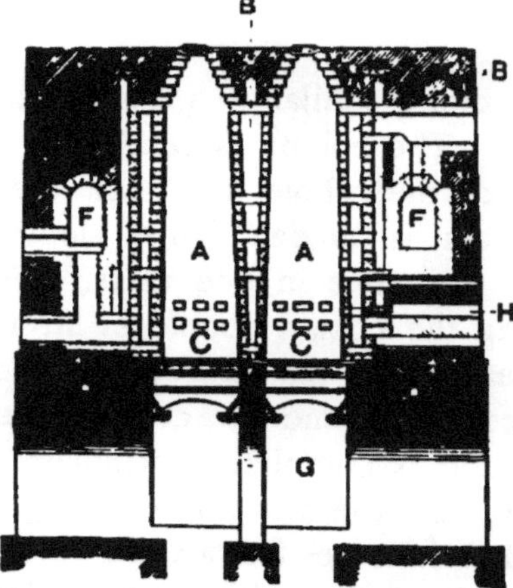

FIG. 27.—Appolt Coke-oven. A, coking chambers; B, combustion space; C, openings for escape of volatile matters into combustion space; F, flues; G, arched spaces under retorts; H, openings to admit air.

covered with coke-dust during coking. The whole block is surrounded by a strong wall containing the flues, of which there are 16, 8 on each side—4 communicating with the top, and 4 with the bottom of the combustion space. These gather into two horizontal side flues, F, which lead to a main flue. Each of the small flues has a damper, so as to regulate the heat through the block.

Passages communicating with the outside, supply air for the combustion of the volatile matters expelled from the coal in the retorts. These pass into the combustion space through openings 5 inches by 2 at various levels, and there meeting with air, are burnt, the heat passing by conduction through the brickwork of the retort.

The chambers are charged from above. Each has a capacity of about $1\frac{1}{2}$ ton of coal. The working is practically continuous, fresh coal being introduced immediately a charge is withdrawn, and the chambers are charged in regular order to keep up the supply of gas.

In these ovens, burning of the coke is avoided, and the yield is greater. The time is shortened by the large amount of heat stored in the mass of the masonry, and the charging of the coal into very hot retorts. On account of their slight build, they are unsuitable for coking coals which swell on heating, owing to their liability to damage from the force necessary to dislodge the coke.

To prevent damage by expansion and contraction of the brickwork, a space filled with sand or loose material is built in the surrounding walls.

The coke produced is of good quality, and in larger quantity, since air cannot find its way into the retorts. The coal is coked in about 24 hours, owing to the large heating surface presented by the retorts. The pressure of the coal above increases the density of the lower portions.

In the **Coppée coke oven** (Fig. 28), the retorts are horizontal arched chambers, open at both ends, and tapering slightly from front to back. They are about 30 feet long, 1 foot 8 inches wide at the back and 1 foot 5 inches in front, and 3 feet 6 inches high. They are closed at each end by two doors, one 3 feet and the other about 1 foot high, luted round to exclude air while coking is going on. A series of vertical flues, V, are built in the side walls of the chamber. These communicate with the coking chamber, and by the passage D with the air. At C they join the horizontal arched flue H running under the chamber from end to end. The gases burn in these flues, the air being heated by its passage through the hot masonry above the ovens, and its supply regulated by the dampers D. A very high temperature is attained.

FIG. 28.

The coal is introduced through openings in the top of the chamber. The ovens are built in blocks of thirty or more, and the flues all merge into one main flue. The distinguishing feature of the Coppée ovens is that they are worked in pairs. It will be observed (Fig. 28) that the flues from both chambers A pass into H. This flue joins H′ by a passage at the back, so that the gases pass backward through H and forward through H′ before passing into the main smoke-flue J at P. In this way, the gases from each aids the coking of the other charge. One retort of the pair is freshly charged when the charge in the other is about half coked and is giving off volatiles rapidly. The surplus heat from the latter, passing under the former, increases the rapidity of coking in the earlier stages, and while the amount of volatiles from the latter diminishes as the coking nears completion, the newly charged one is distilling rapidly, and the surplus heat maintains the temperature at its highest pitch to the end. A more complete combustion is also obtained of the volatiles given off at the beginning of the coking. The coke is pushed out by a ram from the back, and quenched as it leaves the chamber. In coking very bituminous coals in the Coppée, air can be admitted into the chamber if desired. These ovens are largely used in South Wales, yield excellent coke, and are less liable to damage than the Appolt. They are suitable for the treatment of crushed and washed coal.

In the ovens at present considered all the volatile matters have been burnt. As has been shown, these contain many valuable constituents, which, could they be collected without impairing the character of the coke, would form an important

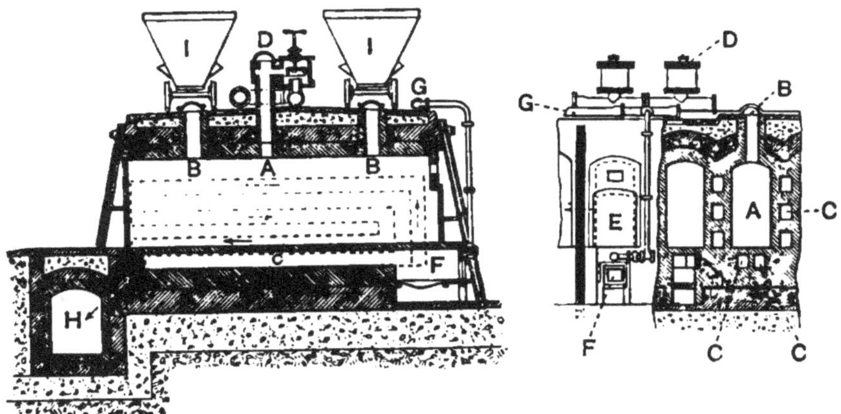

FIG. 29.—Simon-Carves's Coking-oven. A, coking chamber; B, charging openings; C, flues; D, pipe for removing gases, etc.; E, door; F, fireplace; G, gas-main supplying ovens; H, main flue.

source of income. The whole matter rests on a question of temperature, whether the necessary heat can be obtained with sufficient rapidity to produce good coke, after the condensible parts (tar) of the volatile matter and the ammonia have been removed. This problem has been successfully solved by ovens in which the regenerative or recuperative principle is applied.

The **Simon-Carves oven** may be taken as a type of its class. It consists of a rectangular arched chamber (Fig. 29), 23 feet long, 6 feet 6 inches high, and 19½ inches wide, and takes a charge of about 4¼ tons. In the top, at B, are two charging openings, through which the coal is introduced from hopper waggons. These are closed while coking is going on. In the middle of the roof is a 10-inch opening, by which the gases are drawn off through the valve D, and pass into the 10-inch iron gas-main above the battery of ovens. The gases are drawn off by an exhauster, and passed through a series of iron pipes, which are cooled by water, to condense the tar. They next pass through scrubbers and washers, in which the ammonia is dissolved out, and the gases are then led back to the ovens, under which they are burnt. They enter by nozzles into the fireplace E, on the bars of which a thin fire was formerly kept. When air heated by regenerators is supplied this is unnecessary, and the fireplace is abolished.

Under the chamber are two flues, C C′. The products of combustion pass backward along C, and return forward by C′. They then rise by the vertical flue to the highest of the horizontal flues in the side of the chamber, through which they pass down in a zigzag manner, and away into the main flue H. The recuperator for heating the air by means of the waste heat consists of a series of flues, smoke-flues, and air-heating flues, alternating with each other.

The whole of the ovens being at work (on gas), the working is practically continuous, fresh coal being introduced immediately after the removal of the coke. In starting, it is necessary to heat the block of ovens to a coking heat by burning off a few charges in the ordinary manner, without removal of tar, etc. The combustion of the uncondensibles, after this temperature has been reached, is sufficient to maintain it. The ovens are built in blocks, and a high stack produces the necessary draught through the flues. The yield is 15 per cent. greater than the beehive oven, and the coke is of good quality, although not quite so dense and silvery in appearance as is produced in some other ovens. The recuperator ovens work off a charge in 48 hours. The coking is uniform, a high temperature being maintained throughout, and the character of the coke varies little on account of the thin slices in which it is coked. The familiar columnar form of beehive coke is missing.

In the Otto-Hoffman, Bauer, and other ovens, regenerators of chequer work are employed, which require reversal, and consequently greater attention, but, so far as economy of heat is concerned, are more effective.

All ovens of this type are built of most refractory bricks. One of the main difficulties is the burning out of the flues.

Qualities of Coke.—Good coke should be—

(1) dense and compact;
(2) firm, not friable;
(3) uniform in character;
(4) as free from sulphur as possible; and
(5) should have good cell structure;

in order that it may burn freely, and develop great local heat under strong and hot blast, and not crumble and block the air-ways under the pressure of material above. The quality preferred by iron smelters is that which has a silvery appearance, strongly marked.

Sulphur in Coke.—In coking, much of the sulphur in the coal is expelled as carbon bisulphide and sulphuretted hydrogen. Water thrown on red-hot coke causes sulphuretted hydrogen to be generated from sulphides it contains, and any one who has stood near a mass of coke while being quenched will appreciate the offensive smell of this gas which prevails in the vicinity. The addition of salt, carbonate of soda, lime, manganese dioxide, and other bodies, has been made to the coal to be coked, with a view to retain the sulphur as sulphides not decomposed by iron, that is, in a form in which it would not pass into the metal smelted with it. These efforts are unsuccessful from various causes. Quenching has only a superficial effect, as the sulphides are only decomposed by water at red heat. Proposals have also been made to pass superheated steam through the mass while coking. As will be seen (p. 75), at a high temperature, the coke itself decomposes the water, and a less yield is obtained.

Coking of Non-caking Coal.—Coke may be produced from non-caking coal by mixing it with pitch, tar, etc., before coking, or by mixing it with strongly caking coal.

CHAPTER VI.

GAS FUEL.

THREE kinds of gas are employed for heating furnaces.

(*a*) Air gas, or producer gas, made by passing air through a deep layer of carbonaceous matter, whereby carbon monoxide is produced, mixed with nitrogen (from the air), and smaller amounts of other gases, dependent on the fuel used.

(*b*) Water gas, made by passing steam through incandescent carbonaceous matter, whereby the water is decomposed and carbon monoxide and hydrogen generated.

(*c*) Natural gas, consisting mainly of marsh gas.

The advantages of gas over solid fuel are—

1. More complete combustion can be ensured.
2. Better control of the temperature is obtained.
3. Greater uniformity of heating.
4. In regenerative furnaces—in which it is employed—a great saving of fuel takes place, and high temperatures can be more readily attained.
5. Better control is obtained over the atmosphere of the furnace, whether oxidizing or reducing, etc.

Producer Gas.—When a limited supply of air is passed through a deep layer of incandescent carbonaceous matter, the oxygen is converted into carbonic oxide, CO. This, mixed with the nitrogen of the air and small quantities of CO_2 formed, and the products of distillation of the substance employed—hydrogen, hydrocarbons, etc.—constitute producer gas. Moisture entering with the air is decomposed, hydrogen and CO resulting, which latter mixes with and enriches the gas.

By this means the whole of the substance—fixed carbon as well as volatile matters—can be gasified, the only residue being, as in burning, the ash.

The composition of gas thus obtained varies somewhat according to the mode of production and nature of the material used. If the gas be cooled to remove water vapour, even wood sawdust, or any poor fuel, may be employed to produce high temperature by combustion of the gas in regenerative furnaces.

Three types of producers are in use. The original **Siemens producer**, with a grate, in its present modification is shown in Fig. 30.

The fuel is contained in an arched chamber, C, of the form shown, the bottom of which consists of fire-bars. Underneath is an ash-pit, A, closed by the folding doors D, through which the steam-jet blast-pipe P passes. The bottom of the ash-pit is a water-trough, in which the ashes are cooled, and the steam generated passes up into the producer.

The gases pass off by the opening O into a vertical shaft

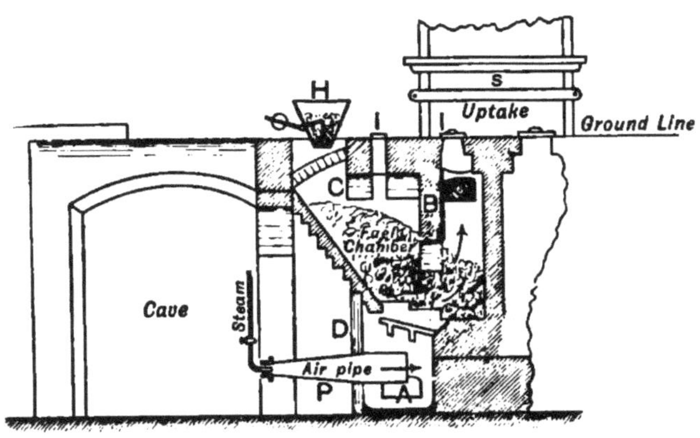

FIG. 30.

S, called the "uptake." H is a hopper, from which fresh fuel is charged; I I are inspection openings, closed while working; and B is a bridge hanging down from the top, so as to prevent any air which may be introduced while charging from forming an explosive mixture by mixing with the gas without passing through the fuel. The top of the hopper is provided with a sliding door, which is shut before lowering the cone, to allow the fuel to descend into the chamber. The bridge also promotes the decomposition of the heavy tars by causing the products of distillation to descend through the heated lower portions. These chambers are usually built in blocks of four, and the uptake for the block is divided into four sections, each of which has a damper, so that any one of the producers may be stopped without interfering with the others.

The **Wilson gas producer** is an example of the cupola type of gas producers, without a grate. It is cylindrical in form, and consists (Fig. 31) of an outer casing of iron plate, lined with refractory brickwork. The fuel is introduced from the hopper at the top, which is provided with a sliding cover. The cone is counterbalanced. The bottom of the producer

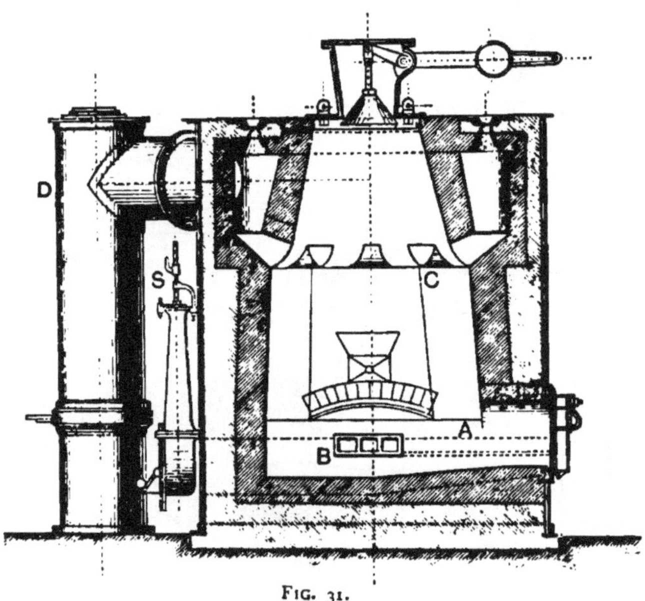

FIG. 31.

is of brickwork. A raised hollow ridge of brickwork crosses the bottom of the chamber.

The air forced in by the steam-jet S blowing into the mouth of the trumpet tube as shown, is delivered into this flue, and enters the chamber by the ports B on either side. Two cleaning doors, A, are provided for the removal of the clinker at intervals. While this is being done, the fuel is supported on iron bars thrust across the chamber through doors provided for that purpose, the steam being meanwhile shut off. In the upper part of the producer is a circular flue, which communicates with the fuel chamber by the openings

C. From this flue the gas is led away by the downtake D to the gas culverts. Openings round the top of the producer permit of the interior being inspected. The chamber is kept full of fuel, and, as the products of distillation must descend through the heated mass before getting away, the tars are largely decomposed.

All steam and water vapour entering a producer is reduced to CO and H ($H_2O + C = CO + H_2$). The oxygen obtained in this way is not mixed with incombustible gases (N), as in the case of air, and the gas is accordingly enriched. But as the same amount of heat is absorbed in reducing the water as is given out in its production, heat will be absorbed, and the producer cooled accordingly. Hence the amount of steam admissible is limited.[1]

To decompose 9 parts of water requires . . . . . . (9 parts of water yield 1 of hydrogen and 8 of oxygen)	34,462 heat units
8 parts of oxygen combining with 6 of carbon to form CO give out 2473 × 6	= 14,838 ,,
Balance lost . . .	19,624 ,,

The quantity of hydrogen converted to CH_4 is practically nil. Some H_2S is generated.

The fuel generally used is washed coal slack, but carbonaceous matter of any kind may be employed. When coal is employed, the distillation products are mixed with the gas. The producers are generally at some distance from the furnaces, and the gas conveyed to them in culverts. In some cases, the gas producer takes the place of the fireplace in a reverberatory furnace, as is the case with the Bicheroux and Boetius furnace.

In Head's new furnace, arrangements are made whereby part of the CO_2 produced by burning the gas is caused to pass through the producer. The CO_2 present is again reduced to CO. There is, therefore, a saving in fuel, the carbon in the CO_2 being used again. Heat is, of course, absorbed in its reduction. This is largely furnished by the excess of heat in the gases when they enter the producer. It would be impossible to return the whole of the CO_2 continuously for regeneration into CO to the producers. The proportion of nitrogen in the gas is unaltered.

It will be observed that in converting solid fuel into gas, part of the heat—that given out by the carbon in burning to

[1] Practical tests show that the best results are obtained by using 5 per cent. of steam.

CO in the producers—is lost unless the gas passes without cooling to the furnace. The great advantages derived from its use, and the waste heat recovered in the regenerators, more than compensate for this loss, and a great saving of fuel is effected where high temperatures are required. For low temperatures, gas fuel is less satisfactory.

Water Gas is a mixture of carbon monoxide and hydrogen, produced by passing steam through incandescent carbonaceous matter.

Natural Gas consists mainly of marsh gas, and is given out in immense quantities in oil regions. It burns with only a faintly luminous flame. It is applied very extensively in Pennsylvania for furnace purposes. The supplies are said to be falling off.

COMPOSITION OF GASEOUS FUELS.

	Coal gas.	Siemens gas.	Wilson gas.	Blast-furnaces gases.	Natural gas.	Water gas.
Carbon monoxide	7·82	24·20	26·44	26·29	2·0	44·4
Carbon dioxide	—	4·20	5·30	10·53	0·8	2·86
Hydrogen	47·6	8·20	11·32	1·96	—	49·61
Marsh gas	41·53	2·20	2·34	2·3	95·75	0·5
Other hydrocarbons	3·05	—	—	—	1·45	—
Nitrogen	—	61·20	54·60	58·92	—	2·53
Percentage of combustible matters	100·0	34·60	40·10	30·55	99·2	94·51

CHAPTER VII.

IRON.

THIS metal is employed in the arts in three forms: as cast iron, wrought iron, and steel of various kinds. Pure iron is a soft, greyish-white metal, very malleable and ductile, and highly tenacious. It is prepared by electrolyzing a solution of iron and ammonium chlorides, sulphates, or oxalate, or by reducing precipitated ferric oxide by heating it in a current of hydrogen. Prepared thus at a low temperature, it takes fire spontaneously in air, but does not if prepared at a high temperature. After fusion, pure iron exhibits a crystalline, scaly fracture. It is softer than wrought iron, and is not affected by heating to redness and quenching in cold water. It is scarcely acted on by sulphuric and hydrochloric acids in the cold, but dissolves on heating. It is highly magnetic, and welds readily. Its specific heat is 0·113, and its specific gravity 7·675. It melts at a lower temperature than platinum—about 1600° C. In mass it is unaffected by dry or moist air, oxygen, or water, if pure and free from carbonic acid gas. In the presence of this body it is readily attacked. At a red heat it is rapidly oxidized in air, forming a scaly coating of oxide. Red-hot iron decomposes water, liberating hydrogen.

$$3Fe + 4H_2O = Fe_3O_4 + 4H_2.$$

When molten, it dissolves or occludes various gases in considerable quantities. Hydrogen, carbon monoxide, and nitrogen are thus taken up and given out on cooling.

The above physical properties are present in a greater or less degree in cast and wrought iron and steel, the extent to which they are modified depending on the purity of the substance.

These bodies consist of iron containing varying proportions of carbon, silicon, manganese, sulphur, and phosphorus, and occasionally copper, arsenic, tungsten, chromium, and other metals.

Iron and Carbon.—The great differences in the properties of cast and wrought iron and steel are mainly due to the presence of carbon in the metal, depending on the *amount* and the *manner* in which it exists in the iron.

The maximum amount of carbon taken up by pure iron is stated by Riley to be 4·75 per cent. In cast iron containing manganese a little over 5 per cent. may be present. Steel may contain up to 1·8 per cent., while the carbon in wrought iron seldom exceeds 0·25, and may fall as low as 0·05.

Carbon may be imparted to iron—

(1) by heating it, embedded in charcoal, at a high temperature for a prolonged period;

(2) by melting iron in contact with carbon, which it dissolves (see Cast Steel);

(3) by the decomposition of carbon monoxide, carbon being deposited and carbonic acid ultimately produced (though not by a simple reaction), as in the blast furnace;

(4) by heating it in contact with gaseous or liquid hydrocarbons, such as paraffins, which are decomposed;

(5) by the decomposition of cyanides, *e.g.* potassium ferrocyanide (yellow prussiate of potash), $K_4FeC_6N_6$, as in case-hardening.

When cast iron cools from fusion, the carbon may remain uniformly distributed through the mass—combined carbon—or a portion of it may separate out in scales resembling graphite. The extent to which separation occurs depends on the rate of cooling and the quality of the metal. Slow cooling, and the presence of silicon and aluminium in the metal, favour the separation, while manganese retards it. When rapidly cooled, nearly all the carbon remains in the combined form. The properties of the iron are modified according to the *amount* and *manner* in which the carbon is held.

Combined Carbon hardens the metal, lowers its melting-point, destroys its malleability and welding power, and tends to make it brittle. The extent to which these effects are produced depends on the amount. In white cast iron, containing as much as 3 per cent., the metal is brittle, breaks with a silvery-white fracture, melts more readily, and passes through a pasty stage in fusing. It is extremely hard, and this property is permanent. In *steel for cutting-instruments*, the amount varies from 0·5 to 1·5 per cent. The hardness and fusibility are increased, the malleability and the welding power diminished in proportion to the amount of carbon present. In this case, however, owing probably to its freedom from other impurities, the degree of hardness can be modified by special treatment: heating to redness and slow cooling rendering the metal soft, while rapid cooling, such as quenching in cold water, etc., renders it hard. The degree of hardness can be modified by subsequently heating it to a lower temperature (see Tempering Steel). When hardened, the metal is brittle. The tensile strength and elasticity of steel are very high. It is magnetized with greater difficulty than pure iron, but *retains* its magnetism.

In wrought iron and mild steels, the carbon exercises an influence in the same direction in proportion to the amount present. Below 0·3 per cent. the metal is not sensibly hardened, even on rapid cooling.

Graphitic Carbon is met with only in cast iron, and occasionally in steel. It reduces the strength of the metal by interposition between the particles, and does not affect the grains of iron themselves. Hence some very grey pig irons are exceedingly soft, and their melting-points very high.[1]

When iron is dissolved in hydrochloric or sulphuric acid, the combined carbon passes off in combination with the hydrogen as foul-smelling compounds, soluble in alkali. Graphitic carbon remains as insoluble. Combined carbon dissolves in nitric acid, giving a brown solution, the depth of colour imparted depending on the amount present (Eggertz Colour Test).

Silicon occurs in cast iron in amounts varying from 0·5 to 12 per cent., being reduced in the furnace. The amount present depends on the working conditions of the furnace —temperature, rate of driving, proportion of fuel, etc. It

[1] The relations existing between carbon and iron—and, in fact, between iron and other elements commonly associated with it—is a problem presenting much difficulty. The generalization given above—combined and free carbon—only expresses part of the truth. When white cast irons, free from manganese, are heated for a prolonged period, at a high temperature, but below fusion, embedded in red hematite, the characteristic brittleness is lost, and the metal becomes more or less malleable (see Malleable Castings). It would appear—since no appreciable diminution in the amount of carbon present takes place—that the carbon contained in the metal separates from it and remains distributed in a finely divided state throughout the mass, *free* but not *crystalline.*

Further, the carbon in hardened steel differs from that in the annealed or unhardened metal. The two states being known as "hardening" and "carbide" carbon respectively. Probably in both the latter cases the carbon is in combination, and both exist in white iron. There are, therefore, *four* conditions in which carbon exists in iron.

Free { (*a*) graphitic, in grey cast iron; (*b*) amorphous (free but non-crystalline), in annealed castings.

Combined { hardening carbon, in hard steel / carbide carbon, in annealed steel } in white pig iron.

renders cast iron more fusible, weak, and brittle. The extent to which it occurs in other forms of iron will depend on the degree of purification. In small quantity, it hardens and weakens the metal. It lowers the melting-point, and its presence in mild steel favours the separation of occluded gases. In cast iron it tends to the separation of the carbon as graphite.

Manganese.—This metal is reduced in the blast furnace. Some pig irons made for special purposes—ferro-manganese—contain up to 85 per cent. of metallic manganese. Pig irons containing more than 7 and less than 20 per cent. are known as "Spiegeleisen" (German = "mirror-iron"), so called from the bright crystalline fracture. With larger percentages, the structure becomes more granular. Ordinary pig iron contains from 0·0 to about 2·5 per cent. Its effect is to whiten the iron by retarding the separation of graphite. Manganese lowers the melting-point, and cast iron containing it does not pass through a pasty state before fusion.

Manganese is looked upon as the principal physician of the steel maker. Iron free or nearly free from carbon, which has been exposed to an oxidizing atmosphere in a fused state at a high temperature, loses its nature and becomes rotten. It has the properties of **burnt iron.** This is probably due to the formation of a suboxide of iron, which is diffused through the mass. Manganese has a greater affinity for oxygen than has iron, and, on its addition, reduces the oxide, forms manganous oxide, and passes into the slag. The iron regains its malleability, etc. The addition made for this purpose always slightly exceeds that required to remove the oxygen, the excess necessary depending on circumstances, notably on the amount of sulphur present. It varies from 0·2 to 0·5. Manganese is consequently found in all mild steels made by the Siemens, Bessemer, and other direct processes. It has also a corrective action on the effects of sulphur.

Pig irons containing manganese are usually freer from sulphur.

Irons containing much manganese lose their magnetic property.

Sulphur is the greatest enemy of the iron and steel maker, on account of its pernicious effects and the difficulty of removal. It combines chemically with iron when heated with it, forming several well-defined sulphides. Ferrous sulphide (FeS), used for preparing sulphuretted hydrogen, and iron pyrites (FeS_2), are the best known. Its presence in malleable iron and steel induces red shortness—that is, the metal cannot be worked at or above red heat, but cracks under the hammer. It renders iron more difficult to weld, and hence the necessity of clean fuel, free from sulphur, for smithy purposes.

The removal of sulphur in purifying pig iron is difficult, and requires that the slag shall be highly basic, and *the fluxes used free from sulphur.* In pig iron its effect is to throw out carbon as a scum, and whiten the iron, making it harder and stronger. Up to 0·3 per cent. it is not objectionable in foundry irons, for castings which do not require fitting and turning, such as columns, etc. Such irons, however, cast indifferently, as they flow sluggishly, and contract on solidifying.

Phosphorus.—This element combines with iron with great readiness, forming phosphides. It is reduced from the phosphates in the charge, in the blast furnace, and taken up by the iron. It renders the metal more fusible and more fluid when molten, and causes it to expand slightly on solidifying. Irons containing it are employed in making fine light ornamental castings. The metal is weaker and more brittle. In ordinary pig iron it is present from 0·0 to 1·5 per cent., depending on the nature of the ore and the fluxes used. Practically all the phosphorus in the charge passes into the metal, unless the slags are very highly basic, and contain a large proportion of oxide of iron, as in the processes for making malleable iron direct from the ore. Its effect on malleable iron and steel is to increase the hardness more rapidly than does carbon. This hardness is not affected by heating and cooling, as is the case with that element. Steel and iron containing it are cold short, and brittle, although they work well when heated. Mild steel should not contain more than 0·08 per cent. The presence of 0·2 to 0·3 per cent. in malleable iron does not sensibly diminish its tenacity or working properties, owing probably to its structure.

Iron and Nickel.—Up to 1·5 per cent. nickel does not produce any increase in the hardness of iron, but increases its toughness and diminishes the tendency to corrosion. Nickel steel is now being used for armour plates.

Chromium up to 1·5 per cent. increases the hardness, tenacity, and ductility, without diminishing the toughness. Hence its employment in shell metal. Ferro-chrome is an alloy of iron and chromium used for introducing it into the steel.

Aluminium is introduced for producing sound castings in mild steel. It is also added to cast iron for foundry purposes, producing finer-grained and stronger castings.

Tin renders iron cold and red short, and unweldable.

Copper in small quantity renders iron red short, and lowers the tenacity.

Tungsten hardens iron and diminishes its malleability. Mushet's steel is an alloy containing from 8 to 9 per cent. of tungsten. It does not require quenching, but is self-hardening—that is, cannot be annealed and rendered soft by prolonged heating. It is brittle, almost silvery white in colour, and very fine grained. Molybdenum is being introduced for the same purpose.

Oxides of Iron.—Three oxides of metallurgical importance are known. **Ferrous oxide** (FeO), **ferric oxide** (Fe_2O_3), and **magnetic oxide of iron** (Fe_3O_4).

Ferrous Oxide (FeO) is not known in the free state. In combination it forms salts as ferrous sulphate (copperas, or green vitriol) and carbonate of iron. It has a great affinity for silica, with which it combines to form a fusible silicate ($2FeO.SiO_2$). This is the principal constituent of many slags produced in refining iron, and in copper and lead smelting. When slags consisting of silicates of iron are heated with carbon, as in the blast furnace, a large proportion of the iron is reduced to the metallic state. The resulting metal is known as "cinder" pig.

Ferric Oxide (Fe_2O_3).—This occurs in a hydrated form (with water) as iron rust, and naturally as various ores of iron. It forms in combination with acids the ferric salts. It has little affinity for silica. If ferrous silicate is roasted in an oxidizing atmosphere, the FeO is largely converted into Fe_2O_3, which separates out. When Fe_2O_3 is strongly heated it gives up oxygen, and is converted into Fe_3O_4. It is reduced to the metallic state by carbon, carbon monoxide, hydrogen, and cyanogen, and oxidizes both silicon and manganese.

Magnetic Oxide of Iron (Fe_3O_4) occurs native as *magnetite*. It is the principal constituent of the scale which forms on red-hot iron when exposed to the air, or when steam is passed over red-hot iron. It is attracted by a magnet. It fuses at almost white heat, and on solidifying forms a bluish-black, crystalline, lustrous mass, and is present to a large extent in "best tap cinder," the slag from furnaces for reheating iron. Its oxidizing power is less than ferric oxide. It is unaffected by exposure, and a layer consequently protects iron from rusting, if the coating is dense and continuous. Iron is, however, electro-positive to it, and if the coating is imperfect, and the iron is exposed, in the presence of moisture, an electrical action is set up which results in the rapid corrosion of the metal.

Barff's Process for protecting iron articles from rust, consists of coating articles with a film of magnetic oxide by bringing them, at a full red heat, into contact with superheated steam. A firmly adherent, dense, but thin coating is thus formed.

Bower's Process.—In this process the coating is formed by heating the articles in a gas furnace, the atmosphere of which is made alternately oxidizing and reducing, by regulating the air-supply. The oxidation produces a thicker but more porous coating, the outer layers of which contain Fe_2O_3. This is afterwards reduced to Fe_3O_4, and the coating consolidated and rendered more adherent by the reduction at the high temperature which prevails.

Ores of Iron.—The principal ores of iron are magnetite, red and brown hematites, specular iron ore, spathose, clay ironstone, and black-band ore.

Magnetite (Fe_3O_4), consists of iron and oxygen. When pure, it contains 72·4 per cent. of metal. It is black or steel-grey in colour, and often crystalline or granular. It gives a black mark on unglazed porcelain (streak), is readily attracted by a magnet, and often magnetized. It constitutes the "lodestone."[1] Its specific gravity is 5·1, and its crystals regular octahedra. It occurs abundantly in Norway, Sweden, United States, Canada, Siberia, etc., in mass.

Magnetic or titaniferous iron sand consists of grains of magnetite with a small quantity of oxide of titanum, derived from the weathering of certain felspathic rocks in which it occurs largely. The lighter matters have been washed away, and the heavy magnetite, with the titanic oxide and other heavy matters present has accumulated. Deposits occur on the shores of Labrador, New Zealand, West Indies, Bay of Naples, etc.

Red Hematite, so called on account of its red colour and streak, consists of ferric oxide (Fe_2O_3), and contains, when pure, 70 per cent. of iron. It occurs both in dense and earthy forms. Kidney iron ore is a dense variety which occurs in masses with a rounded exterior. The specific gravity is 5. It is usually very pure, containing only silica (quartz) as impurity. The more earthy forms of the ore are less pure. The soft varieties are used for fettling puddling furnaces under the name of "puddlers' mine." It occurs in Cumberland (round Whitehaven), Lancashire (Ulverstone), Glamorganshire, and Shropshire, etc., Canada, United States, Spain and Algeria, Saxony, Bohemia, and the Hartz mountains.

Specular Iron Ore is crystallized ferric oxide, and has the same composition as red hematite. It has a steely-grey colour, often darker on the surface, and iridescent. The crystals are modified rhombohedra, often, as in the black incrustations on hematite, thin plates. The streak is red, and the specific gravity 5·2. "Micaceous iron ore," and "iron glance," are names given to a variety with a grey metallic lustre which readily separates into thin plates or scales. Some varieties are ground up for paint, on account of their high density. This ore occurs in Devonshire, Elba (the mine has been worked for 2000 years), Russia, Spain, Nova Scotia, and elsewhere.

Brown Hematite.—Brown iron ore, Limonite—includes a series of substances consisting of *hydrated* ferric oxide—ferric

[1] Anglo-Saxon *lædan*, "to lead."

oxide and water, chemically combined. It contains, when pure, 60 per cent. of iron.

Brown hematite proper is a heavy dense form, sometimes with radiating structure and shining exterior like kidney ore. It is generally very pure. *Gothite* is of an iron-black colour, with crystalline structure. *Wood hematite* resembles wood in being made up of alternate light and dark concentric layers. *Bog-iron ore* is a light, porous, dark-brown mass, often very impure. *Lake ore* is obtained in Sweden and Finland from the bottoms of shallow lakes by dredging with a net. *Umber* is a dark brown, light, earthy body, containing often manganese, copper, and cobalt. *Yellow ochre* is so called from its yellow colour; it is soft, earthy, and unctuous. All varieties give a yellow or brownish streak. The purity of the ores varies greatly. The Forest of Dean ore from the coal measures contains 89 per cent. Fe_2O_3 and 10 per cent. of water, and yields exceptionally pure iron.

The brown hematites of the North of Spain, resulting from the decomposition by atmospheric influences of veins of spathic ore, are very pure, and often contain much manganese. They are imported for making manganiferous pig iron, and for use in steel-making. The Northamptonshire and Lincolnshire ores are a light yellow and earthy, often full of fossil shells from the oolite. Bog-iron ore yields iron only fit for foundry purposes, owing to the large amount of sulphur and phosphorus contained. The moisture present in brown hematites varies from 9 to 14 per cent. In France, Germany, Spain, and Canada, the principal ores smelted are of this nature. Deposits occur in Devon, Glamorgan, Northampton, Lincoln, Cumberland (Alston Moor), and Durham, India, etc.

Spathose (spathic or sparry iron ore), so called on account of its sparry appearance, consists of crystallized ferrous carbonate, $FeCO_3$ (ferrous oxide combined with carbonic acid). When pure it is of an ashen-grey colour, and contains 48 per cent. of iron. The streak is white. It generally contains more or less carbonate of lime, magnesia, and manganese, which crystallize in the same form, and is often more or less decomposed by weathering, with the formation of hydrated ferric oxide, which colours it brown. Some samples contain 50 per cent. of carbonate of manganese, and it was from these ores that manganiferous pig iron was first produced. The manganiferous brown ores of the North of Spain have been produced by weathering from spathic ore. They occur in Somerset, Durham, Cornwall, Isle of Man, Styria, Carinthia, Westphalia, Prussia, etc.

Clay Iron Stone includes all ores of a compact, earthy, stony character, varying in colour from light grey to brown.

They consist of ferrous carbonate mixed with more or less clayey matter. Sometimes the deposit is nearly pure carbonate of iron, but in an uncrystallized state. The brown colour is due to partial decomposition with the formation of the hydrated oxide (brown hematite). This ore, which is the most important British ore, occurs (1) in nodules, sometimes very large, made up of successive layers, in clay; and (2) in beds, in the coal measures and oolitic strata. The iron present varies from 20 to 37 per cent. The ore is of low specific gravity and of a stony appearance, but on calcining becomes black owing to the formation of Fe_3O_4. Lime, magnesia, and manganese, as carbonates, etc., iron pyrites, galena, zinc blende, and copper pyrites, as also phosphates and sulphates, principally of lime, as well as clay, frequently accompany clay iron stone, rendering the pig iron smelted from it, as a rule, less pure than from other ores. The sulphur in such iron varies, but rarely exceeds 0·2 per cent. in "all mine pig."[1] The phosphorus ranges from 0·2 up to 1·5. These ores are worked in South Staffordshire, Derby, Notts, Leicestershire, Warwickshire, North and South Wales, Cleveland district in New York, etc.

Their occurrence in conjunction with coal, limestone, and fire-clay, furnishing all necessary materials for smelting on the spot, has been one of the principal factors in the development of the British iron trade. Similar formations occur in Belgium and Silesia.

Black-band Ore is a variety of clay ironstone, admixed with more or less coaly matter. This sometimes occurs in

[1] A term used to designate iron made from ore without any admixture of cinder from puddling and other processes.

layers, giving the ore a banded appearance, hence the name; it is sometimes present in such large quantity as to colour the ore black, the amount varying up to 30 per cent. These ores occur in North Staffordshire, in Lanarkshire, and in Prussia, etc. They contain from 17 to 30 per cent. of iron. Owing to the bituminous matter present, it is often unnecessary to add fuel in calcining the ore.

Iron Pyrites (FeS_2).—The heavy, yellow, metallic substance so frequently found as "brasses" in coal, occurs extensively. It must be regarded rather as an ore of sulphur than of iron, being used for the manufacture of vitriol. The residues from certain varieties, after burning off the sulphur, and after treatment for the copper they contain, are used as fettling for the puddling furnace, under the name of "blue billy." It consists of ferric oxide (Fe_2O_3).

CHAPTER VIII.

IRON SMELTING.

Introduction.—As already explained (p. 82), when oxides of iron are heated with reducing agents, such as carbon (C), carbon monoxide (CO), hydrogen (H), cyanogen (CN), the oxygen is removed, and metallic iron results. This reaction occurs at all temperatures above redness. Apparently, therefore, the production of malleable iron would be a simple matter, were it not for the facts—first, that the iron itself is so difficult to melt, and second, the infusibility of the associated earthy matters, while, if the temperature is raised, the iron takes up carbon (see p. 78), and at the same time silicon and phosphorus are reduced, and pass into the metal, depriving it of all its malleability and other useful properties.

It follows, therefore, that, to produce malleable iron from ore direct, a low temperature must prevail, the ores must be rich and fairly pure, and the earthy matters fluxed off by some body which will give a readily fusible slag. The only substance available for this is oxide of iron itself. This removes the impurities, mainly silica, as silicate of iron, and by its excess prevents the iron with which it is in contact from taking up carbon. It is obvious that by such methods only the better classes of ore can be treated, and that the reduction is only partial. This method would consequently be very wasteful, while the production would be very limited as to quantity. Methods of this character are still followed in India, Africa, and elsewhere (see p. 118), and formerly were the only methods practised. The carbon, silicon, and phosphorus which enter the iron when the reduction is effected at a high temperature, can be removed by heating the impure metal—cast iron—in an oxidizing atmosphere, or with oxide of iron itself, whereby it is converted into malleable iron. At the higher temperature employed in making cast iron, other substances, such as lime and magnesia, may take the place of the oxide of iron in fluxing off the impurity, and the reduction is complete. Poorer and less pure ores can thus be treated, and as the length of time which the iron remains in contact with the carbon, etc., during reduction is not limited, the furnaces employed may be of any size compatible with efficiency, and the output thus enormously increased.

This indirect production of malleable iron, by first obtaining pig or cast iron, and purifying it, is found to be more economical under ordinary conditions than the direct processes, and is the one generally followed.

The *ore*, after the necessary preparation, is charged, together with the *fuel* (charcoal, coal, or coke), which also serves as the reducing agent, and the *flux* into a tall blast furnace, which is kept full, and the materials, as they melt, sink, and make way for fresh additions at the top. The iron is reduced, and by taking up carbon, silicon, etc., becomes fusible at the furnace temperature, and, melting down, accumulates at the bottom. It is removed from time to time by making an opening into the furnace, and allowing it to run out. This "tap-hole" is at other times kept stopped with a mixture of clay and sand. The slag, after reaching a certain height, flows continuously from the furnace, and is disposed of in various ways (see p. 113).

Preparation of Iron Ores.— The objects aimed at are (1) to remove extraneous matters completely; (2) to break down the ore to pieces of such size that the reduction shall be complete before it reaches that part of the furnace in which the charge is melted down, otherwise oxide of iron would pass into the slag. (3) In the case of spathic ores and clay ironstones, it is desirable to convert the protoxide of iron present into peroxide, to prevent the passage of the iron into the slag, by its combination with the silica in the charge, at a red heat before reduction has been effected.

Washing.—Clay, sand, and similar admixed and adherent matters are removed from heavy ores by washing on iron grids under a stream of water, and stirring the ore about with rakes.

Calcination.—This is by far the most important process to which iron ores are subject prior to introduction into the furnace. It consists (see p. 18) of heating the ore with free access of air.

In this country, only clay ironstones and spathic ores are generally calcined; hematites and magnetites are smelted without this treatment. As they already consist of peroxides of iron, and would lose nothing but some 6 to 12 per cent. of water, which is expelled in the upper part of the furnace, little advantage is derived, and fuel consumed for this purpose would be practically wasted.

In Sweden, however, where these ores are often washed, and where, owing to the lower temperature of blast employed, all the gas collected from the furnace is not required to heat the air, these ores are also calcined, the waste gases being employed for this purpose.

The operation is conducted in open heaps or in kilns of special construction, in which less fuel is necessary and the air supply and temperature can be better regulated.

Calcination in Heaps.—The ore is piled up on a thin layer of coal, the large blocks at the bottom and the smaller stuff above, and covered with the smaller ore. In calcining clay iron stone, some 10 or 12 per cent. of small coal is mixed with the ore; but with black-band ores this is unnecessary, the burning off of the bituminous matter present furnishing the necessary heat. The heaps are about 5 feet high, and the sides slope at about 60°, and are partly covered with small ore. They are ignited at one end, at the base, and allowed to gradually burn through, small refuse being used to check the combustion at any point where it is progressing too rapidly.

Calcining in heaps is wasteful in fuel and heat, and the product is not uniformly calcined. Some parts of the heap will be almost fused, and in black-band ores partly reduced, while others are not calcined through, and require a second treatment.

To overcome these difficulties, kilns of some kind are now generally adopted for all but black-band ores, fired either with solid fuel or waste gas.

Calcining kilns are open-topped, circular, or rectangular structures of masonry or boiler-plate, lined with fire-bricks, with openings near the base for admitting air and withdrawing the material. They are charged from above, and generally work continuously.

Rectangular kilns of masonry, with sloping sides, and lined with fire-brick, are employed for calcining in South Wales.

Gjer's calciner, in use in the Cleveland district, is shown in Fig. 32. It consists externally of boiler-plate, and is lined with fire-brick. It is supported by a cast-iron ring resting on short pillars. The descending ore is diverted outwards by the cast-iron cone which projects upward into the kiln. It is charged from the top, the coal and ore being brought in trucks on rails. The ore, on removal, is despatched to the furnace.

Fillafer's kilns, used in Styria and Carinthia, for the treatment of spathic ores, are narrow rectangular chambers 9 feet high, 4 feet 8 inches long, and 2 feet wide, with fire-bars at the bottom, and a space beneath. Waste gases from the blast furnace are admitted from flues in the side walls near the bottom, and burn in air drawn through the grate-bars. The roasted ore is discharged by withdrawing one or more bars, and allowing it to fall into the space below.

In calcining, water and carbonic acid gas are driven off, and some of the sulphur in the pyrites present. The bituminous matter in black-band ore is also burnt off. This leaves the ore in an open, fissured, porous condition, in which it is readily acted on in the furnace. The peroxidation of the iron has already been noted. Certain pyritous ores before calcining are exposed to the air, or **weathered**, for long periods. The sulphides of iron and copper in the ore are converted into sulphates. These are washed out by the rain or by drenching the heap with water. By this means much sulphur is got rid

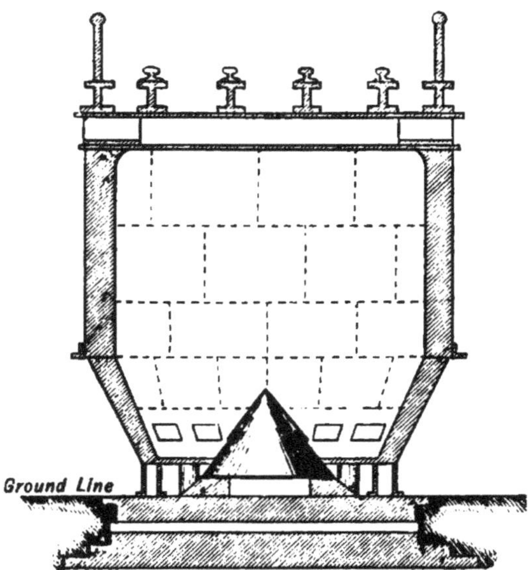

FIG. 32.—Gjer's Calciner.

of, and a purer iron results on smelting. Weathering also facilitates the detachment of pieces of adherent rock.

The *size* of the pieces of ore introduced into the furnace depends on the nature of the ore, the fuel, and the rate of reduction. The slower the descent of the material, and the more open the character of the ore, the larger the pieces may be. Magnetites and hematites are broken in from 1 to 2 inch cubes. The others may be in larger pieces. Stone or ore breakers are used for this purpose.

The Blast Furnace used in iron smelting has undergone

great structural changes in the last thirty years. The massive masonry structures, braced with iron, formerly employed have given way to the lighter type of furnace known as the cupola blast furnace, and whereas formerly the top of the furnace was open, and the gases were allowed to freely escape and burn at the top, they are now usually closed, and the gases, which resemble producer gas in composition, are collected and conducted by iron pipes to the ground, where they are burnt for heating the blast and for raising steam.

The *height* and *capacity* of furnaces has also greatly increased, so that a "make" of 1000 tons of pig iron per week per furnace is not infrequent.

A furnace of this type is shown in Fig. 33. It will be observed that it consists externally of a boiler-plate casing lined with refractory material, the upper part being supported on columns. The furnace increases in diameter from the *throat* downwards until a maximum diameter is attained at the *bosches*, and then contracts more rapidly, until at a point somewhat above the tuyere openings it becomes nearly cylindrical, a form which it preserves to the bottom. This form, which is the result of gradual development, has the following advantages:—The gradual increase in diameter in passing downward permits of the better admixture of the materials as they work outwards in descending, and by the increased volume detain the ore for a proportionately longer period in this part of the furnace, until certain reactions are completed. In the lower part of the furnace, the rapidity with which the fuel is being consumed, and the materials fused up, with the consequent great contraction in bulk, necessitate the rapid narrowing of the furnace chamber, in order that the charge shall descend to the hearth with regularity. The exact internal form, notably the height of the bosches, and the diameter at this point as compared with the height of the furnace, depend on the nature of the ore and fuel, and on the class of iron produced in the furnace.

The casing of the furnace is made of $\frac{3}{8}$ to $\frac{1}{2}$ inch boiler-plate, well riveted together, and the lining is constructed of 5-inch fire-brick blocks, chisel dressed on both faces and joints, so as to ensure perfect setting and uniformity in the outline of the furnace. These may only form some 18 inches of the lining, and be backed with ordinary fire-bricks, or the

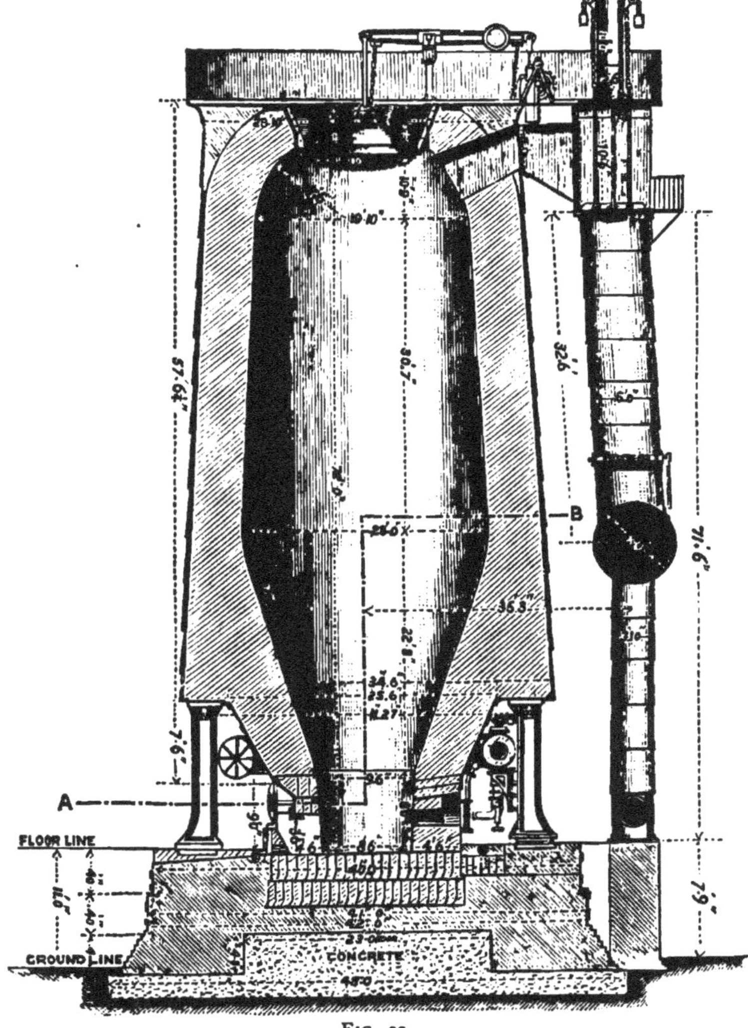

FIG. 33.

whole thickness (3 feet 6 inches to 5 feet) of lining may be thus made.

The upper part of the structure is carried on columns,

which rest on a stone kerb bound with iron bands, enclosing the hearth of fire-brick blocks. This stone foundation rests on a bed of concrete. On the top of the columns rests a cast-iron ring, some five inches thick, cast in segments, and on this the superstructure is built. The lower part of the furnace is supported by plating also attached to the pillars.

From the tuyeres downwards the hearth is supported by iron bands, and in various other ways.

The height of such furnaces varies from 60 to 100 feet,

FIG. 34.—Lower part of furnace. A, blast-main (horseshoe); C, tuyere; E, fore hearth; F, crucible.

and the diameter of the bosches from 17 to 30 feet. The ratio of height to diameter $= \frac{H}{D}$ varies from $2\frac{1}{2}$ to 5, the prevailing proportion being from $2\frac{1}{2}$ to $3\frac{1}{2}$; but excellent results have recently been obtained from furnaces in which the ratio was increased. Such a furnace, with a height of 85 feet, and bosch diameter of 17 feet, has recently been put to work. The height of the columns varies from 10 to 18 feet 6 inches, the diameter of the hearth from 8 to 9 feet, and its depth from the tuyeres to the bottom 3 feet 6 inches. The number of tuyeres ranges from 4 to 7, and they are distributed at equal distances round the hearth.

Fig. 34 shows an enlarged view of the lower part of the furnace. The blast is brought to the blast-main A, which is an iron pipe, lined internally with fire-brick, if hot blast is employed, supported on stanchions at a height of about 7 to 8 feet. It encircles the furnace except in front, and from this at regular intervals the blast is conveyed to the tuyeres by vertical iron pipes passing downwards, each of which is provided with a throttle-valve to regulate the air-supply. Suspended from this by hangers is the "goose-neck" B, which articulates with the vertical main by a ball-and-socket joint. At the bend of B a mica plate is inserted, known as the "furnace eye," and through it the working of the furnace may be inspected. A sheet-iron blow-pipe, which slides telescopically on B, conveys the blast through the tuyere-block, C, into the furnace. These tuyeres enter the furnace through small arched openings known as tuyere houses. On their lower side they rest on fire-brick wedges on the top of the hearth, the remainder of the space being filled up with fire-bricks, and closely luted. They generally project a little beyond the front line of the furnace, and are cooled by the circulation of cold water supplied from the water main which surrounds the furnace. When, as occasionally happens, the nose of a tuyere is burnt off, it is removed, and a new one inserted, blast being stopped for the time being.

The use of water-tuyeres became necessary with the introduction of hot blast, on account of the increase in temperature in this region. With cold blast, the absorption of heat by the expansion of the cold air so cooled the furnace just in front of the tuyeres, that the slag solidified on the end, and formed a prolongation or slag nose, from the length of which the working could be judged.

Fig. 35 is the **Staffordshire tuyere.** It is a hollow conical iron box encircling the blow-pipe, cooled by water circulating between the casings.

In the **Scotch tuyere**, the water circulates round a coil of wrought-iron pipe embedded in a cast-iron block.

The **Spray tuyere** is a hollow casing, and the water is sprayed on the front through holes in the pipe conveying it. The *size* of tuyere is proportioned according to the volume of air and the pressure employed, so that the blast is carried well into the furnace, and does not creep up the sides. The blow-pipes from the goose-necks fit tightly into the tuyere, and are luted round with clay, to prevent escape of air.

The *hearth*, or *crucible*, of the furnace, in which the iron collects, as shown in Fig. 34, sometimes projects forward beyond the vertical, forming what is known as the "fore hearth," E. This is closed by a "dam," supported by a water-cooled iron plate, "dam-plate." In the top of the dam is a groove, the "slag-notch," through which the slags flow continuously, after reaching that level, and through which the blast blows to keep it clear. The structure above the fore hearth is supported by a water-cooled cast-iron girder, "tymp iron," and the space between the dam and the upper part of the furnace is closed by an iron plate (Fig. 34) backed with fire-brick. In many furnaces the fore hearth is abolished, and the slag flows out through a slagging-hole in the furnace wall.

The *tap-hole* is at the side of the dam, at the bottom.

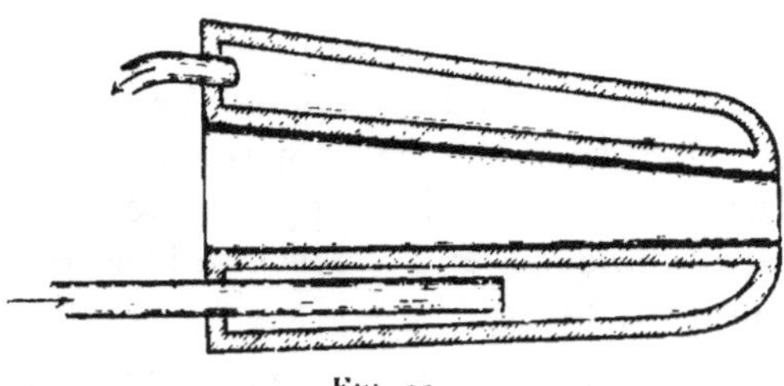

FIG. 35.

It is a rectangular opening about 15 inches by 2, kept closed until the metal has accumulated almost to the level of the slag-notch, by a mixture of clay with sand, or coal-dust. This is broken away by a pointed bar when the furnace is *tapped*, and the metal allowed to flow out. While this is being done the blast is shut off.

The throat of the furnace shown in Fig. 33 is closed by a **cup-and-cone** arrangement. A truncated hollow iron cone cast in pieces and bolted together, supported in the throat of the furnace, and resting on the masonry by a broad flange, forms the "cup"—the opening into the furnace being closed by the "cone" supported from one end of a lever projecting over the top. At the other end is a counterpoising weight, slightly heavier than the cone, and the apparatus for controlling its motion, when it is lowered to allow of the descent of the material (previously charged into the cup) into the furnace.

Another method of closing the top of the furnace is sometimes adopted. The cone is replaced by a conical iron funnel, through the stem of which the gases pass by a short vertical tube into a cross-tube supported above the furnace, and thence to the down-comer. The end of the vertical tube dips into water in a circular trough round the stem of the funnel, thus ensuring a gas-tight joint. In this arrangement the gases are removed from the central part of the furnace throat. Explosion doors are arranged on the end of the cross-tube and elsewhere, to prevent damage from possible explosions, arising from air admitted during charging. They consist of a simple lifting flap, which falls back into its place when the force has spent itself.

These methods of closing the throat, when properly designed, give the most regular distribution of materials attainable in mechanical charging, the ore, flux, and fuel falling in an annular heap some distance from the wall and from the centre. The larger stuff consequently has an equal tendency to roll towards the sides and middle, and, the inequalities working away as the charge descends into the wider part of the furnace, uniform obstruction is offered to the blast, which in consequence distributes itself regularly, having no tendency to creep up the sides or middle, as would be the case if the large stuff accumulated there. In the Staffordshire and Barrow districts, the tops are often not completely closed. Sometimes a tube some 5 feet long and 7 feet wide is hung in the throat. This is kept full of material, which acts as a sort of stopper. The gases are withdrawn by a flue connecting with the annular space between the tube and the furnace lining.

A *gallery* or *platform* surrounds the throat of the furnace. This is covered with iron plates, and slopes slightly towards the edge of the cup, which stands some 3 or 4 inches above it, and acts as a stop for the barrows.

The **gases** pass through the flue at the side into the wide iron pipe "down-comer" (Fig. 33), and are led away to boilers, stoves, etc., where they are burnt. The excess of gas burns at the mouth of the standpipe seen on the right of Fig. 33. Communication with the gas main is cut off when the cone is lowered to introduce the charge.

Chambers in which dust is deposited, and apparatus for the removal of ammonia and tar from the gases of furnaces using coal as fuel, are interposed in many cases.

Lifts.—The charge, consisting of ore, flux, and fuel, is raised to the top by means of steam, hydraulic, or pneumatic lifts, or sometimes is drawn up inclined planes. Occasionally the situation, near a hillside, permits of the trucks themselves being drawn up by a locomotive.

In the **water-balance lift**, two cages or platforms sliding between guides are connected by a steel rope passing over a pulley at the top. The bottom of each cage is a water-tank. Its capacity is such that, when filled, the cage, of which it is a part, together with the empty barrows, is somewhat heavier than the other cage with *full* barrows on, but the tank *empty*. By filling the tanks at the top of the furnace, and emptying them by valves at the bottom, the material can be elevated by gravitation. The water required may be taken from a neighbouring hillside stream, or pumped up for the purpose, as at Dowlais.

In Sir Wm. Armstrong's hydraulic lifts, the movement of a ram, connected with multiplying gear, through from 5 to 7 feet, raises the charge to the top.

In pneumatic lifts, a counterbalanced cage is attached to a piston working in a large cylinder, so that an air-pressure of a few pounds applied above or below, gives the necessary motion, up or down.

Steam lifts generally consist of a small winding-engine. On the drum, two ropes passing over pulleys at the top, and attached to cages sliding between guides, are so arranged that as one cage ascends the other descends.

The Charge.—The proportions of the materials in the charge must be separately determined for each ore and fuel, and even for each furnace, the fuel consumption being influenced by the volume, temperature, and pressure of the blast, as well as by the nature of the fuel. More coal, for example, must be used than when coke is employed. With clay ironstones containing, after calcination, 35 to 42 per cent. of iron, the charge consists of from 48 to 57 cwts. of ore, 19 to 25 cwts. of coke, and 10 to 14 cwts. of limestone per ton of iron made, the temperature of blast varying from 500° to 700° C., and its pressure from $3\frac{1}{2}$ to 5 lbs. In furnaces using coal, from 2 to $2\frac{1}{2}$ tons replace the coke.[1]

[1] The coal is coked in the upper part of the furnace.

For red hematite the charge consists of 33 to 40 cwts. of ore, containing from 50 to 60 per cent. of iron, 7 to 10 cwts. of limestone, and 19 to 25 cwts. of coke, and, if very siliceous, about $1\frac{1}{4}$ cwt. of aluminous ore (see Fluxes).

In smelting magnetites with charcoal, from 16 to 25 cwts. of charcoal are consumed per ton of iron made.

Some ores contain all the ingredients necessary for fluxes, and are described as *self-going* or self-fluxing.

The proportions existing between the ore and fuel in the furnace are described as the **burden.** It is "light" when the fuel is in large proportion, and "heavy" when the quantity of fuel is diminished.

The height of the materials in the furnace is maintained at a constant level—the stock line—fresh additions being made at intervals of from 10 to 20 minutes.

The Blast.—Time was when bellows worked by hand supplied the air necessary for the low furnaces—seldom exceeding 10 feet high—then employed. Some of our large modern furnaces require as much as 50,000 cubic feet of air per minute, at a pressure of $3\frac{1}{2}$ to 5 lbs. This is supplied by means of blowing-engines, some of which are capable of delivering as much as 60,000 cubic feet of air per minute. These engines are of various forms. A huge cylinder, sometimes 12 feet in diameter and 12-feet stroke, fitted with a solid piston, is provided with valves in such a manner that, as the piston travels to and fro, air is drawn in at one end and expelled from the other, or, in other words, the cylinder is double acting—filling one end and discharging the other, in whichever way the piston is travelling. These blowing-cylinders are connected with steam-engines, by which they are worked. The air is delivered at a pressure varying from a few ounces up to 9 lbs., if desired. Low pressures are employed for charcoal furnaces, and the higher in coke and anthracite furnaces. From 3 to $4\frac{1}{2}$ or 5 lbs. are the pressures commonly employed. About $2\frac{1}{2}$ to 3 lbs. is employed in furnaces using coal.

NOTE.—In the same furnace, with similar materials, the pressure of the blast influences the quality of the iron produced. Low pressures and large volume, and consequent rapid driving, increase the make but lower the grade of iron. Higher pressures, with less volume and slower driving, diminish the make but improve the quality (see Reactions in Furnace).

Hot Blast.—Formerly the air was supplied to the furnace at the temperature of the atmosphere. In 1828, Neilson, at the Clyde Iron Works, commenced the use of heated air, and in a few years its use became general.

The advantages derived are—

(1) The use of raw coal (of certain classes) instead of coke.

(2) Much less fuel is required in the furnace, owing to the heat carried in by the air.

(3) The temperature in front of the tuyeres is increased, and the fusion zone of the furnace brought lower down.

(4) The furnace works with greater regularity, and is more under control, not being affected by atmospheric influences.

The temperature of blast employed depends on the fuel and class of iron made. With charcoal it is only heated to from 200° to 350° C.; with anthracite and coke, temperatures of from 700° to 830° C. are employed. The higher temperatures tend to produce greyer irons, containing more carbon and silicon.

Blast Stoves.—The air is heated by passing it through cast-iron pipes, or through brickwork regenerators, heated by the burning of the waste gases collected from the top of the furnace.

Fig. 36 shows a cast-iron pipe stove. The air circulates through the pipes in the chamber, from end to end, as shown.

In pipe stoves a temperature of 550° C.[1] (1022° F.) cannot be exceeded without danger of rapid oxidation and fracture of the pipes. The products of combustion escape at a temperature equal to that to which the air is raised, and carry off at least one-half of the heat generated.

In regenerative hot-blast stoves, the principle of the Siemens regenerative furnace is embodied. The waste gases collected at the throat of the furnace are burnt in the

[1] This is seldom realized in practice. From 600° to 900° F. are the usual temperatures.

stove, and the products of combustion drawn through brickwork flues on their way to the stack. The heat is thus absorbed, and by passing the blast through the stove in a direction opposite to that taken by products of combustion, it may attain the temperature of the stove. At least two such stoves working alternately — one being heated up while the other is in use for heating the blast—will be required. The advantages derived from their use are : (1) higher temperature of blast; (2) less loss of heat, resulting in greater fuel economy; and (3), absence of difficulties arising from the burning, cracking, and leakage of iron pipes.

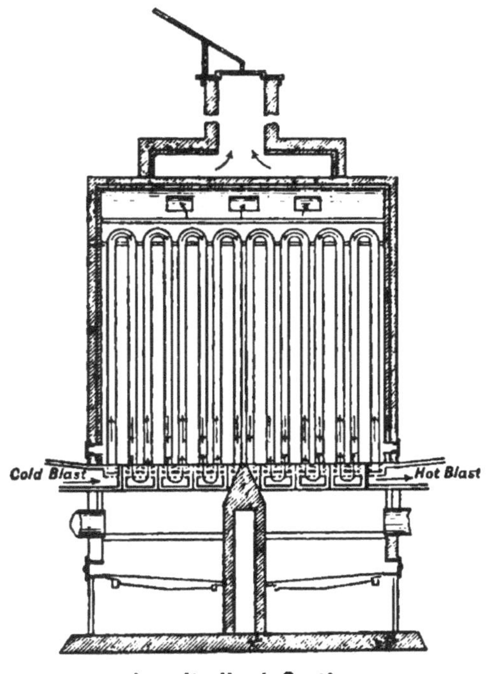

Longitudinal Section.

FIG. 36.—Cast-Iron Pipe Stove.

Cowper's Regenerative Hot-blast Stove is shown in Fig. 37.

The waste gases from the furnace are brought by the culvert V, and enter the flame or combustion flue O, through the valve F. Here they are burnt by the admission of a suitable supply of air through G. The products of combustion ascend and are drawn down through the passages in the brickwork P—ordinary chequers, or built of special bricks—in passing through which they are deprived of nearly the whole of their heat. The brickwork gets first heated near the top, but the heating gradually extends downwards. The chequer is carried on cast-iron grids supported by short brick columns. Doors for cleaning-out purposes communicate with the space beneath. In passing through the chequers, the products of combustion are cooled down to about 150° to

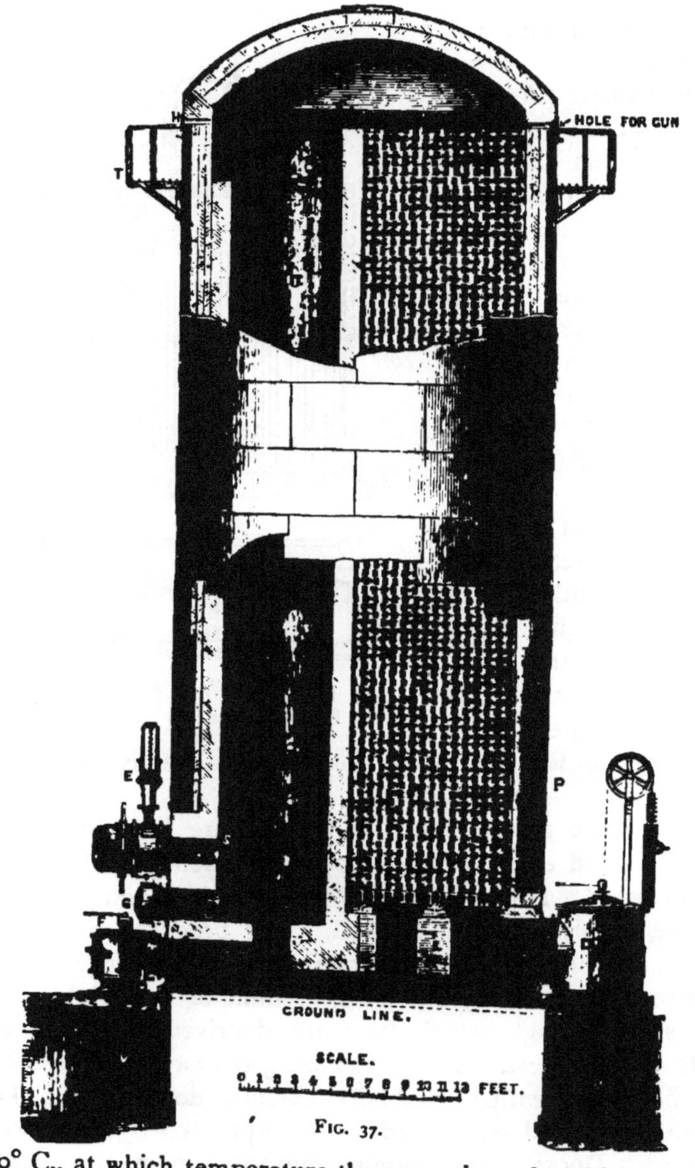

Fig. 37.

200° C., at which temperature they pass into the chimney-flue

U, and thence to the stack; the heat thus carried away doing useful work by creating a draught through the stove.

When the stove has been heated to a maximum about halfway down, the supply of gas is stopped, and the air- and chimney-valves closed. The valve of the cold-blast main, which enters the space under the chequer, and the hot-blast main E, which communicates with the combustion-flue, are opened. The cold air rising through the hot brickwork gradually becomes heated *by contact* with it as it ascends, until it has attained the temperature of the stove. It then passes through the remaining upper portion without further absorption of heat, and, being collected at the top, passes down the flame-flue into the hot-blast main.

To prevent, as far as possible, dust being carried into the stoves, the gases are passed through dust-boxes, in which the current of gas is slowed down and the dust in large measure deposited.

Whitwell's Stove is shown in Fig. 38. The chequer-work of the Cowper is replaced by vertical walls, to facilitate cleaning.

In regenerative stoves the blast is heated to temperatures varying from 1200° to 1400° F., and the stoves are changed at intervals of from half an hour to two hours. The large capacity of these stoves renders a regulating air-vessel to equalize the pressure, between the blowing engine and the furnace, unnecessary.

In **blowing in** a furnace, as in heating up any large mass of brick-work, the greatest care has to be exercised. The masonry is first dried by wood fires, and then fuel gradually added until the furnace is half full; a small blast, through, say, a ¼-inch nozzle, is then introduced, and a little limestone added to flux off fuel ash. The charging of material may then commence, a much larger proportion of fuel than will ultimately be used being present in the charge. The size of the nozzles is gradually increased until the full volume and pressure of blast have been attained, the time elapsing before this is reached sometimes being as long as 18 days. The ore and flux are also gradually increased until they reach the normal amount.

In "blowing out" a furnace, the burden is gradually diminished, and at the last only fuel and a little limestone are fed in, so that the furnace is completely cleared.

When an accumulation of material forms at any point in the furnace, which from some cause or other will not work down, the irregularity is described as "scaffolding," or "hanging," of the charge.

The sudden descent of the "hanging" material, owing to the melting away of the support, is called a "slip." It is sometimes attended with serious consequences.

The formation of infusible masses of iron, often containing titanium,

FIG. 38.—Whitwell Hot-Blast Stove. A, gas-valve; B, hot-blast main; C, chimney valve; D, cold-blast main; E, doors for removing dust; F, cleaning holes through which scrapers are introduced; G, air inlets; P, inspection openings.

in the hearth and lower part of the furnace seldom occurs since the introduction of highly heated blast. They are known as "bears."

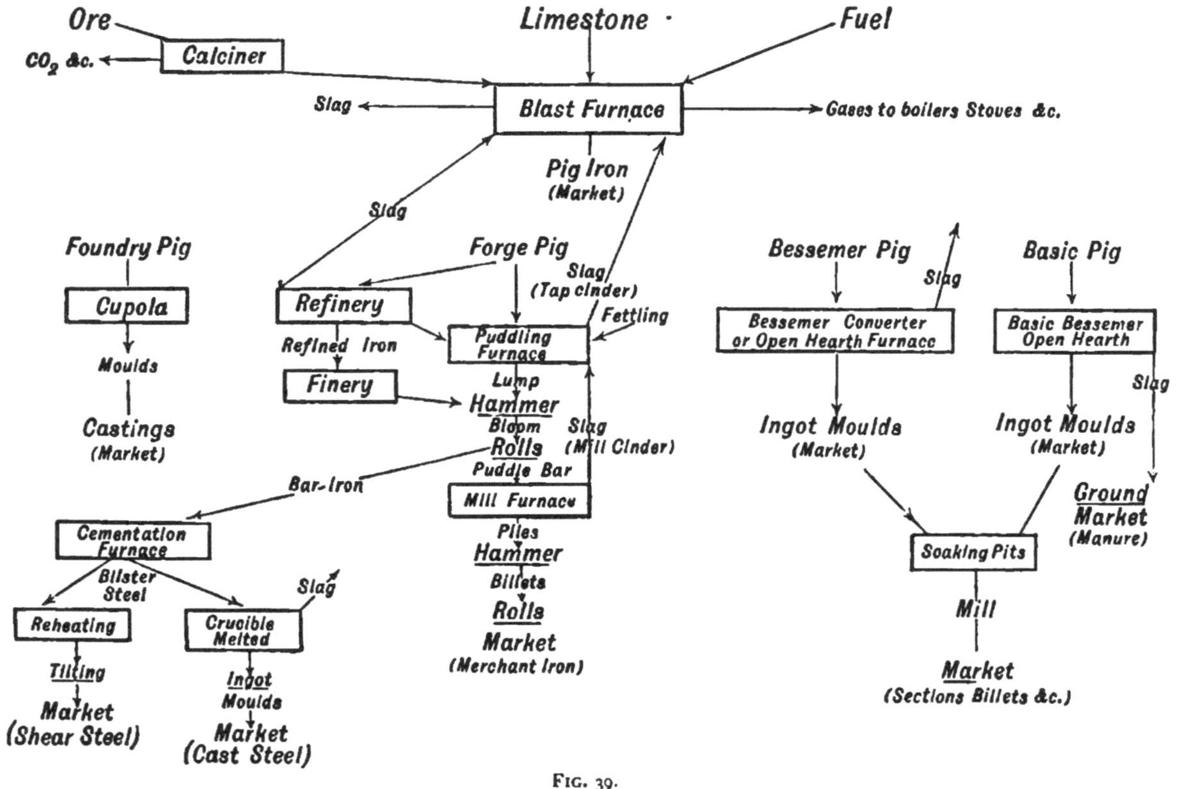
Ore
CO_2 &c.
Calciner
Limestone
Fuel
Slag
Blast Furnace
Gases to boilers Stoves &c.
Pig Iron
(Market)
Slag
Foundry Pig
Cupola
Moulds
Castings
(Market)
Forge Pig
Refinery
Refined Iron
Finery
Slag
(Tap cinder)
Fettling
Puddling Furnace
Lump
Hammer
Bloom
Rolls
Slag
(Mill Cinder)
Puddle Bar
Mill Furnace
Piles
Hammer
Billets
Rolls
Market
(Merchant Iron)
Bar-Iron
Cementation Furnace
Blister Steel
Reheating
Tilting
Market
(Shear Steel)
Slag
Crucible Melted
Ingot
Moulds
Market
(Cast Steel)
Bessemer Pig
Slag
Bessemer Converter or Open Hearth Furnace
Ingot Moulds
(Market)
Basic Pig
Basic Bessemer Open Hearth
Slag
Ingot Moulds
(Market)
Ground
Market
(Manure)
Soaking Pits
Mill
Market
(Sections Billets &c.)

FIG. 39.

The furnace being in full work, the charging of fresh material at the top goes on at regular intervals. The metal accumulates until the hearth is full, and is then tapped out by piercing the clay stopping of the tap-hole with a pointed bar. A channel leads from the front of the furnace to a sand bed which slopes very slightly from the furnace. This main channel is continued, in the sand, to the front of the bed, sending out side channels at intervals, which serve as feeders for long rows of roughly made U-shaped open moulds, lying like huge combs in the sand. The feeders are picturesquely named "sows," and the metal in the moulds "pigs." The row of moulds furthest from the furnace is first filled, and the others in succession, the metal being prevented from entering the upper rows by stops. When the metal becomes solid, the pigs are detached, and the sows broken up into convenient lengths with sledge-hammers.

The pigs are about 4 feet long and 4 inches across the surface.

Swedish pig and ferro-manganese are generally cast in wide, open, iron moulds into a plate, which is broken up.

CHAPTER IX.

CHEMICAL REACTIONS OF BLAST FURNACES.

In considering the chemical changes going on in the furnace, the conditions under which it works must be borne in mind.

The materials charged in at the top occupy a considerable time in descending the furnace, varying from nine hours to two or three days, according to the nature of the material, the quality of the iron being produced, and the quantity of blast; the charge descends most rapidly with the heavy burdens and large blast employed in making white iron. In its descent, the oxide of iron is reduced to the metallic state by the carbon monoxide in the ascending current of hot gases, and the various substances found in pig iron are introduced into the metal. But little consumption of fuel takes place until it arrives in the vicinity of the tuyeres. Here the oxygen of the air combining with the carbon produces carbon monoxide (and perhaps a little carbon dioxide, which is at once reduced to monoxide by the excess of carbon present), producing heat for the fusion of the metal and slags.

With hot blast this zone of fusion is immediately above the point at which the blast enters. With cold blast, owing to the absorption of heat by the expansion of the cold air blown in, this part of the furnace is somewhat cooled, and the zone of most rapid combustion extends higher up the furnace.

This gas, together with the nitrogen of the air and any hydrogen resulting from the decomposition of water-vapour in the air blown in, ascends, and is the principal active reducing and carburizing agent in the furnace. As the materials in the lower part of the furnace are burnt away and melted up, those above gradually descend, passing through hotter and hotter regions till the fusion zone is reached.

The reactions by which the iron is reduced and carburized are of a somewhat complicated nature, depending on the relative affinities of carbon and iron for oxygen, at varying temperatures.

The only change taking place in the upper part is the gradual heating up of the charge. When a sufficiently high temperature has been attained, the reduction of the iron commences. The carbon monoxide, CO, combines with the oxygen in the oxide of iron, forming carbon dioxide, CO_2, and liberating the iron—

$$Fe_3O_4 + 4CO = 3Fe + 4CO_2$$

This action commences at temperatures considerably below a red heat, and the oxide is gradually reduced to a spongy mass of metallic iron, which includes all the gangue in the ore.

At temperatures a little below, and above redness, spongy iron decomposes carbon monoxide, carbon being deposited and oxide of iron formed, which is subsequently reduced by carbon—

$$3Fe + 4CO = Fe_3O_4 + 4C$$
$$Fe_3O_4 + 2C = 3Fe + 2CO_2$$

These reducing and carburizing actions go on side by side.

The spongy iron, with its deposited carbon in its further descent, is subject to the oxidizing and reducing influences of the carbonic oxide and carbon dioxide, and in the middle region these almost balance each other, so that little change takes place so far as the iron is concerned. In the lower part of the furnace, any residual iron oxide is reduced, probably by cyanides present, and the metal fuses, dissolving up part of the carbon deposited in it, together with silicon, manganese, and phosphorus, which have been reduced in its descent.

The limestone charged in as a flux is reduced to lime, CO_2 being expelled, in the upper part of the furnace when the charge has attained a sufficient degree of heat. When fusion occurs, it combines with the gangue, and produces a slag.

The **silicon** in the pig is reduced from silica (SiO_2) in the charge, in the lower and hotter regions of the furnace, by the joint action of carbon and iron; for while carbon does not reduce silica alone, it does so at high temperatures in the presence of iron. The amount reduced depends on the temperature and the rate of descent.

Manganese is only reduced in the blast furnace by the direct action of carbon at high temperatures. Carbon monoxide only reduces oxides of manganese to the lower oxide, MnO. The reduced metal alloys with the iron.

Phosphorus is introduced into the iron by the reduction of phosphates in the charge, by carbon in the presence of silica at high temperatures. Practically, the whole of the phosphorus in the charge enters the metal—

$$2Ca_3(PO_4)_2 + 3SiO_2 + 10C = 3(2CaOSiO_2) + P_4 + 10CO$$

Sulphur is introduced in another manner. The sulphide of iron existing in the coke and other materials in the furnace liquates into the iron during fusion, owing to its specific gravity being greater than that of the slag.

By the use of a large proportion of lime, this may to some extent be prevented, and the sulphur carried into the slag. This is probably owing to the liberation of *calcium* by the *silicon* in the iron, which combines with the sulphur, forming calcium sulphide.

From a consideration of the above, it will be seen that, to produce a highly carburized iron, time must be given for the metal to take up carbon, and a higher temperature will be required in its fusion. These conditions favour also the reduction of silicon, and, in order to keep that element low, more lime must be employed. Hence "grey" irons are freer from sulphur than "white" irons.

Alkaline cyanides are produced and accumulate in the lower part of the furnace by a series of complicated reactions, from traces of alkali existing in the charge, and materially assist in the reduction of the last portions of iron oxide.

The first formation of the cyanide may be thus represented—

$$K_2CO_3 + 2C = 3CO + K_2$$
$$K + C + N = KCN$$

NOTE.—It should be noted that the lime serves a twofold purpose. It combines with the silica and other gangue in the ore, and at the same time prevents the silica from combining with oxide of iron, and forming a scouring slag (see p. 114). The high temperature which prevails ensures the liquefaction of the more difficultly fusible slag thus produced.

THE PRODUCTS OF THE BLAST FURNACE.

These are (1) pig iron; (2) slag; (3) furnace gases.

Pig Iron.—Pig, or cast iron, is classed as grey, mottled, or white, according to the appearance of the fractured surface. **Grey Pig Iron** has a crystalline or granular appearance, a dark iron-grey colour, is soft and easily turned, chipped, or filed. The carbon, of which it contains a large proportion, has mainly separated as graphite.

Such irons require a higher temperature to melt them than white iron, but are more fluid when molten, and expand slightly in solidifying. They are specially suitable for foundry purposes. They are weaker than whiter iron, but less brittle. The strength increases as the grain gets finer. They usually contain more silicon and less sulphur than white iron. Grey pig iron dissolves less gas than white, and consequently casts more soundly. Greyness is not a proof of the absence of phosphorus and other objectionable impurities.

They are classified as Nos. 1, 2, 3, etc., according to greyness, No. 1 being the greyest.

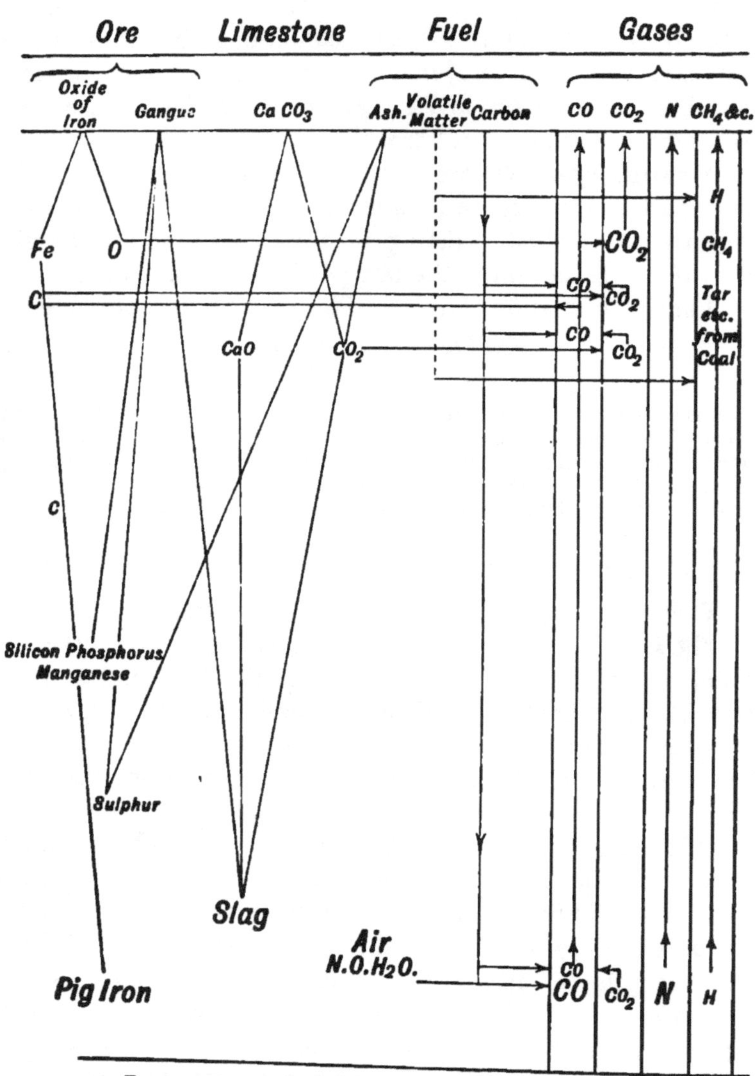

FIG. 40.—Diagram showing chemical changes in blast-furnace.

Mottled Iron presents the appearance of a matrix of white iron with grey spots. Its carbon is present in both the free and combined state. The quantities in the respective states are about equal in most varieties.

White Cast Iron presents a white, close, and sometimes crystalline appearance. Its carbon is mainly in the combined form (see p. 78). It is extremely hard and brittle, and usually contains more sulphur and less silicon than grey iron. It melts more readily, but flows more sluggishly than grey iron, giving off sparks in abundance. On this account it is less suitable for making castings. It contracts slightly on solidifying. Most varieties pass through a pasty state before fusing,[1] in which condition, the oxides of iron and slags formed in the puddling furnace can be more readily incorporated with the metal, and the impurities it contains oxidized out with less waste. On this account, malleable iron was formerly made exclusively from irons of this class, either produced *directly* in the blast furnace, if the ores were sufficiently pure, or, if not, grey iron was made in the blast furnace, and afterwards whitened (see Refining, p. 122).

When molten grey pig iron is rapidly cooled, it is rendered white. On this account Swedish pig, which is cast in thin plates in iron moulds, is often white top and bottom, with a grey interior. The surfaces of grey pigs are often white from the same cause (see Chilled Castings, p. 116). From the larger proportion of sulphur usually present in white iron from the furnace, they are classed as **low-grade** irons.

The specific gravity of white iron is greater than that of grey. White has a specific gravity of 7·5 as compared with 7·1.

Grey Forge Pig is a class of iron containing less silicon than other grey irons. The amounts of sulphur and phosphorus must be small. It is of fine grain.

Pig irons for trade purposes are usually classed as Foundry, Forge, Bessemer, and Basic pig. Other descriptions indicating the source from which they have been obtained are also used, *e.g.* Hematite iron, etc.

Bessemer pig iron must be practically free from phosphorus, while Basic pig contains that element in considerable quantities.

Cold-blast pig, owing to the lower temperature prevailing during its production, contains less silicon than the hot-blast pig of the same

[1] All except those containing manganese.

greyness. As this element greatly diminishes the strength of cast iron, strong castings generally contain an admixture of cold-blast pig. Swedish pig, smelted with charcoal, is also used for the same reason.

Spiegeleisen and **Ferro-manganese** are the names applied to varieties of pig iron containing notable quantities of manganese employed for carburization in making mild steel. Spiegeleisen (German = "mirror iron") is so called from the brilliancy of its fractured surface. It is highly crystalline, and breaks with broad, flat, lustrous faces, tinged somewhat with yellow. Up to about 10 per cent. of manganese this becomes more pronounced, but as the percentage increases, the size of the crystals diminishes, and in those containing much manganese, the fracture, though yellowish, is granular. Pig containing from about 7 to 20 per cent. is classed as spiegel., and from 20 to 85 or 88 per cent as ferro. (the common contractions of their names). Pig, containing less manganese than would constitute a spiegel, is made for conversion into steel by the basic open-hearth process. The object is to ensure the greater freedom of the iron from sulphur.

Manganiferous irons are made from spathic ores, and Spanish hematites containing oxide of manganese, in the blast furnace. The conditions necessary are slow reduction, high temperature, and basic slag. To secure this, light burdens, small blast at high pressure and temperature, densest coke, and much flux are employed.

In making ferro-manganese, the slag often contains as much as 13 per cent. of manganous oxide, and is green in colour. This is necessary to give fluidity to the slag. The make of a furnace working on ferro. is little more than one-fifth of its output on grey forge iron. Manganiferous pig contains about 5 per cent. of carbon.

Siliconeisen and **silico-manganese** are irons containing silicon, or silicon and manganese, but practically free from sulphur, etc. From 12 to 21 per cent. of silicon may be present. They are employed in steel manufacture.

Glazy pig is a whitish crystallo-granular iron, in structure somewhat resembling grey iron, but is brighter and whiter. It contains up to 12 per cent. of silicon.

Small proportions of aluminium, chromium, copper, and calcium frequently occur in pig iron.

ANALYSES OF PIG IRON.

	Grey.			Mottled.	White.
	Hematite No. 1 (Greenwood).	Cold-blast Bowling (Abel).	Hot-blast Derbyshire ore.	(Bodemann.)	Frodingham (Author).
Graphitic carbon .	3·045	2·99	3·35	1·59	—
Combined carbon .	0·704	—	—	2·78	2·98
Silicon	2·003	0·97	1·27	0·71	0·96
Manganese . . .	0·309	—	1·01	—	0·505
Phosphorus . . .	0·037	0·5	1·09	1·23	1·41
Sulphur	0·008	0·05	0·02	—	0·28
Iron	93·800		93·26	93·29	93·865
	99·906		100·000	100·000	100·000

Blast-furnace Slag.—As already noted, the slag flows over the top of the dam, or through the slag notch. It is disposed of in various ways. Sometimes it is led into circular cavities in the ground, where it accumulates and solidifies, the block being then lifted by a crane on to a bogie, and dragged off to the tip or cinder-heap.

The more common way of removing it is to lead the molten stream into slag-tubs, where it solidifies. These tubs are waggons with movable iron sides, and are rectangular or conical in form. They are run on rails up to the furnace front, and drawn away by a locomotive when filled. After solidifying, the sides are removed and the huge blocks tipped off.

In Styria, the slag is disposed of in a novel manner. It is allowed to accumulate in the furnace along with the iron, on the top of which it floats. When tapping takes place, the iron forming the lowest layer in the furnace comes out first. As soon as the slag begins to flow, it is diverted into a side channel, at the end of which it meets a stream of cold water. The sudden cooling reduces it to coarse sand, which the velocity of the water carries forward into one of the many rapid streams of the district, and thus gets rid of it.

The slag usually consists of double mono-silicate of lime and alumina, with more or less magnesia, oxide of manganese, and other bases. The general composition is given below :—

Silica	40 to 47 per cent.
Alumina	5 „ 25 „
Lime	30 „ 40 „
Magnesia	1 „ 8 „
Manganous oxide	1 „ 3 „
Ferrous oxide	1 „ 2 „
Soda	traces „ 1·25
Potash	„ „ 2 „
Phosphoric acid	traces only
Sulphur	traces to 2 „

the *general formula* being $3(2MOSiO_2) + 2M_2O_3SiO_2$.

The presence of manganous and ferrous oxides renders the slag more readily fusible. If lime or alumina is in excess, the fusibility is diminished. Magnesia is not such a good flux as lime.

In the above, normal slags only have been considered. In making spiegel, more manganese is present, and sometimes, owing to derangements in working or in the making of white low-grade iron, the oxide of iron rises to as much as 8 per cent. Such a slag is known as a *scouring* slag; it is black in colour, and fuses readily. Its effect is to render the iron whiter by the reduction of the oxide in the slag by the carbon and silicon in the metal.

Blast furnace slags vary in colour from nearly white, through shades of green, blue, or brown, to black.

The green tinge is due to the presence of ferrous oxide. Blue may be due to alumina or alkaline sulphides. The brown colour is ascribed to manganese sulphide? If much ferrous oxide is present, its colour is bottle green or black. Excess of lime makes the slag light coloured and stony.

The character of the same slag varies with the mode of cooling. Rapid cooling makes it glassy; if cooled slowly, it may be stony; while, if gas is escaping while molten, it will be light and porous.

The higher temperature employed in making grey iron permits of the use of larger quantities of limestone in the furnace charge. This causes the slag to be lighter in colour—a light greyish slag almost invariably accompanying the production of grey iron. In consequence of the excess of lime, the sulphur in such slags is higher than in slags produced in making white iron, so that grey forge pig is superior in this respect to white forge pig made from the same ore. The absorption of moisture from the air by the excess of lime present often causes these slags to fall to pieces on exposure.

Slags are utilized to some extent in various ways according to their nature. Certain siliceous slags can be made into blocks for paving, by running the slag into iron moulds, and removing them while still hot into an annealing oven and treating them like glass.

Highly basic slags free from iron are used for making inferior glass.

Excellent concrete bricks, etc., can be made by grinding down basic slags and mixing with about 10 per cent. of milk of lime and moulding the blocks by compression. In course of time they harden like stone. For making bricks, the slag is granulated by water, and for concrete, broken slag mixed with slag sand is employed. If very basic, no addition of lime is necessary. Good cement has also been made from slag. Slag sand is also used in place of ordinary sand for mortar.

Slag wool, made by blowing air through molten slag as it flows from the furnace, is a good non-conducting material. Its use as a steam packing, however, has not been attended with success.

The question of utilizing slag is of great importance For each ton of iron, from 10 to 30 cwts. of slag are produced. This amounts in the aggregate to many thousands of tons yearly, which by its accumulation encumbers the land.

Blast-furnace Gases, as taken off at the throat of the furnace, consist of a mixture of—

Carbon monoxide	25 to 29 per cent.
Carbonic acid gas	6 ,, 11 ,,
Nitrogen	54 ,, 57 ,,
Hydrogen	0 ,, 7 ,,
Marsh gas	0 ,, 3 ,,

In furnaces using charcoal and coke, hydrogen and marsh gas are low, being derived almost entirely from the moisture in the air blown in. Ammonia and tarry matters are also present in the gases from furnaces burning coal. At some iron works these are recovered before burning the gases.

It will be observed that the composition of these gases is similar to producer gas, with a considerable increase in the amount of CO_2 present. The furnace may, in fact, be considered as a huge gas-producer. The excess of CO_2 is accounted for by the reduction of the oxide of iron going on in the upper part of the charge, at a temperature too low for it to be converted into carbon monoxide by the carbon with which it is in contact.

The volume of gas is enormous. Each ton of coal burnt yields nearly 4 tons of gas, measuring 130,000 cubic feet or more according to the temperature.

Kish is a separation of graphite which sometimes occurs while grey irons are cooling and solidifying.

Iron Founding.

For the purpose of making castings, iron is melted down in small blast furnaces, called "cupolas." A very satisfactory type is shown in Fig. 12. The outer iron shell is lined with fire-brick up to the level of the charging hole E, and stands on a raised platform, at a convenient height for filling the ladles with the metal from the tap-hole. The bottom is of fire-brick, and carefully covered with ganister, or sand and clay, and made to slope towards the tap-hole situated in front. At the back of the furnace, at the base, is a movable plate, kept in its place by an iron bar, which passes across it and through two lugs on either side of the furnace. This is for the removal of the residues from the furnace when melting is completed. In some modern cupolas, the furnace is supported on pillars, and the bottom is removable. The ratio of height to diameter is about as 5 : 1 or 6 : 1. The blast is brought by the pipe B to the jacket C which encircles the outer casing, and from which openings, DD, at regular intervals through the lining and casing, admit the air into the furnace. A fire is first made in the cupola, which is then nearly half filled with coke. When fairly ignited, the back plate is fixed and the blast turned on. The metal is charged in pieces weighing about 28 lbs., in layers alternating with layers of coke. A little limestone is usually added to flux off the ashes of the fuel. As soon as metal makes its appearance at the tapping-hole, it is stopped with clay. As the iron melts, it collects at the bottom, and is tapped off into ladles as required. From 1 to 2 cwts. of coke are used per ton of iron.

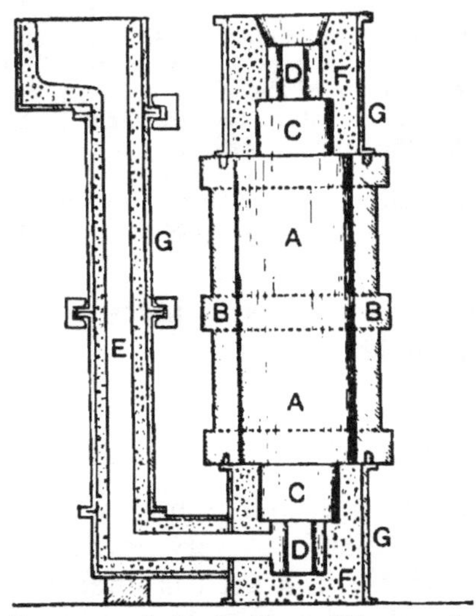

FIG. 41.—A, body of roll; B, chill; C, neck; D, wobbler end; G, boxes; E, running-gate; F, sand mould.

The fuel used in the cupola should be as free from sulphur as possible. Sulphur absorbed from the fuel in melting has a tendency to throw out the carbon and whiten the iron.

As cast iron flows from the furnace, it throws off sparks or "jumpers." This happens to a less extent with grey than with white irons. No. 1 scarcely scintillates at all.

The moulds are made in "green-sand," or a mixture of green-sand with about 8 per cent. of coal-dust, or in "loam." They are well vented, to permit of the free escape of the gas given off.

As before noted, grey irons are best suited for foundry purposes (see p. 109). If chilled on the surface, a thin skin of hard white iron is produced. To prevent this, ordinary sand moulds are blacked. Charred oak wood is the best blacking for light work. Graphite (blacklead) is also largely employed. For heavy work, the heat contained in the body of metal prevents the face from being chilled.

Chilled Castings.—This hardening is made use of in the production of castings of which some part is subjected to wear. The wearing surface, *e.g.* the tread of a car wheel, is rendered hard by chilling the metal. This is effected by using a carefully prepared *iron* mould for the part which is required to be hard, the rest of the article being moulded in sand in which the "chill" is embedded. Rolls for rolling iron, zinc, etc., are thus hardened on the face (see Fig. 41). This practice leaves the body of the casting soft and tough, only the wearing face being rendered hard. The turning, etc., of such castings can only be done by specially prepared tools.

Malleable cast iron and malleable castings are made by packing the articles in red hematite in pots or boxes, from which air is completely excluded, and heating them for a prolonged period.

The articles are thus rendered soft and to some extent malleable. A thin strip thus treated, may be bent double, but generally breaks on attempting to straighten it. The strength of the articles is greatly increased. Castings for annealing are made in special mixtures of pure irons free from manganese, having a tendency to whiteness. The reason for the change during annealing is somewhat obscure. It has been supposed that carbon is removed by the oxide of iron in which they are embedded, but from analysis such does not appear to be the case to any great extent.

The prolonged heating probably causes the separation of the combined carbon in a finely divided state, but differing from graphite, throughout the whole mass, leaving the iron itself purer and more malleable.

CHAPTER X.

MALLEABLE OR WROUGHT IRON.

UNDER this heading are included all classes of iron which can be hammered and forged at red heat, and are not hardened by heating to redness and plunging in cold water. It is usual to further restrict the term to the metal as obtained in a pasty, unfused state, produced either in a direct manner from the ore, or indirectly from pig iron by puddling or analagous processes.

Direct Processes.—As previously noted (p. 87), iron oxides are reduced by carbon or carbon monoxide at dull-red heat, and the earthy impurities can be removed, and carburization prevented, by allowing oxide of iron to pass into the slag. Such a slag is readily fusible, and by hammering the pasty mass obtained, the particles of malleable iron become welded together, and the slag is expelled. All the iron produced by the ancients was thus obtained, and many processes of a similar character are yet in use in India, Burmah, Africa, and other places, where modern civilization and its methods have not yet penetrated.

Various modern "direct" processes for the treatment of special materials, and with a view to the saving of fuel, have also been introduced. Methods of the crudest and of the most refined nature belong to this category.

In Burmah, a hole in the side of a clay bank, some 10 feet deep and 2 feet wide, does duty for a furnace. The front of the bank is strengthened by small branches interlaced and supported by stakes driven in the ground. At the bottom, an opening about a foot high and the width of the furnace is cut through the bank for the removal of the lump of metal and the slag. This is stopped with clay. A row of clay tubes, about 4 inches long, made by plastering clay on bamboo, and afterwards drying and burning them, is introduced about halfway up the opening. These supply the air to the furnace, which works entirely with natural draught. A fire is lighted, and a quantity of charcoal thrown in. The rest of the furnace is filled up with alternate layers of ore and charcoal, and the furnace is left pretty much to itself. In a few hours, slag makes its appearance at the bottom, and is tapped out and examined. If free from shots of iron, it is thrown away. When the furnace has burnt out, the clay breast is broken down and the lump dragged out. It consists of metal, fragments of unburnt charcoal, and slag, and weighs about 90 lbs. It is broken up and sorted, according to fracture, into soft, and hard or steely iron.

In India, most native iron-makers use blast, and the urnaces are generally built above ground. They range in size from a chimney-pot to about 10 feet high, and the blast is supplied by curious contrivances. Bellows made of goat and bullock skins stripped off whole, single-acting wooden blowing-cylinders, the pistons of which are stuffed with feathers, and bellows not unlike ordinary smithy bellows, being employed. In some furnaces, the front is broken down in order to remove the iron; but in others the metal is hauled out by tongs from the top, a fresh charge being at once introduced.[1]

Similar processes are in use in Central Africa.

Easily reducible brown hematites containing over 50 per cent. of iron are the ores generally employed. Little more than half the metal is reduced, the rest passing into the slag. The presence of so much oxide of iron in the slag prevents the reduced metal from taking up carbon, and, in many cases, the low temperature which prevails is unfavourable to carburization. The silica and phosphorus in the ore are removed in combination with oxide of iron in the slag.

FIG. 42.

The famous **Catalan, Elba,** and **Corsican** processes are very similar in character. They still survive to a small extent.

The forge (Fig. 42) in which the reduction is effected is a rectangular hearth, one side of which curves outwards at the top. It is about 21 inches long, 19 wide, and 17 deep. The bottom is a movable block of granite. The tuyere side is built of malleable iron blocks, as also is the sloping side facing the tuyere; the back is of masonry, and lined with fire-clay. The front consists of thick iron plates, placed edge to edge, the lower end being on the ground. The tuyere is of copper, and the blast-pipe lies in it loosely. It is inclined so that the blast strikes downwards. A tap-hole at the bottom serves to remove slag, and to introduce a bar to lift up the mass of iron when the process is completed.

[1] Modern blast furnaces are now in use in India.

The hearth, hot from previous working, is filled with charcoal to the tuyere, and a gentle blast turned on. When fairly alight, a broad shovel is placed a little in front of the tuyere, so as to divide the hearth into two unequal parts.[1] The tuyere side is filled with charcoal, which is moistened. On the other side the charcoal is rammed down tightly, and the space filled with roasted and broken ore, from which the fine stuff has been riddled. This is then covered with a mixture of fine ore and charcoal-dust, and finally with moistened charcoal. The blast is then turned on, and in a few minutes the flame of carbon monoxide makes its appearance at the top. From time to time ore and charcoal are added, being pushed down the sloping side of the furnace into the hearth, under the tuyere, where the reduced iron accumulates. The carbon is repeatedly damped to prevent its too rapid combustion. The slag is removed at intervals and examined. The operation is complete in five or six hours. The lump collected at the bottom is raised in front of the tuyere for a few minutes to raise it to a full heat and melt out slag, and is then dragged from the furnace, and hammered to expel slag. It weighs about 3 cwts.

The mass, which is never homogeneous, is broken up and sorted. The pieces are reheated in the corner of the hearth formed by the tuyere side and back, during the progress of a subsequent operation, and drawn down into bars.

The reduction is effected principally by the CO, formed by the air passing through the carbon before coming into contact with the ore, consequent on the method of filling the hearth, and the downward direction given to the blast. The siliceous impurities are fluxed off by ferrous oxide formed by partial reduction of the ferric oxide in the ore. A very fluid, fusible slag is thus obtained, carburization at the same time being prevented. This is also favoured by the low temperature employed.

The blast is provided by a blowing-machine known as the *trompo*, and the pressure varies from $\frac{1}{2}$ to $1\frac{1}{2}$ lb.

The American bloomery is in somewhat extensive use in

[1] After charging this is removed.

Canada and the United States, and New Zealand more especially, for the treatment of the titaniferous iron sands, ores, etc.[1]

It is a small rectangular hearth with sloping sides, and measures 27 to 28 inches, by 30 to 32 inches along the sides, and at the back is about 33 inches deep. The sides are made of thick cast-iron plates, and the bottom of a water-cooled hollow casting. A single, water-cooled tuyere, inclined so that the blast strikes the centre of the hearth, is introduced at the back about 12 inches above the bottom; the tuyere opening is segmental in form, 1½ inch high and ¾ inch wide. The front of the furnace is 16 inches deep, and at this height is a flat iron plate 18 inches wide. The tapping-hole is situated below this, and at the side of the hearth.

The furnace is worked by filling the hearth with charcoal, and scattering the fine ore over the top of the fire at intervals, additions of fuel being made from time to time. The reduction takes place as the ore passes down in front of the tuyere, but the iron does not fuse. The grains of reduced metal agglomerate in the bottom of the hearth, forming a mass or loup, which is subsequently raised in front of the tuyere, and when heated to a full welding heat, hammered to expel the slag.

The slag is similar to that obtained in the Catalan forge, and the chemical reactions identical.

The blast is usually heated to about 300° C., by circulating through iron pipes in a brick chamber built above the furnace, heated by the waste heat. A pressure of about 1½ lb. is employed.

Only rich ores containing 50 per cent. of iron or over can be economically treated.

One furnace produces about a ton of blooms per day of twenty-four hours, the loup being removed every three hours. They work continuously for a certain part of the year.

Indirect Methods.—Pig iron is converted into malleable iron by removing the silicon, carbon, manganese, and phosphorus by oxidation. During this treatment, part of the sulphur is also removed if the slag produced is highly basic.

The above-mentioned bodies have a greater affinity for oxygen than iron has, and hence, on exposing the metal during fusion and while melted to a blast or current of air, they are oxidized, together with a portion of the iron itself, which forms such a large proportion of the whole.

The silica (SiO_2) phosphoric anhydride (P_2O_5), manganous oxide (MnO), and oxide of iron formed, unite together, producing a fusible slag consisting of silicate and phosphate of iron, with the excess of iron oxide. The carbon passes off in the gaseous state as CO or CO_2.

Instead of employing an *air blast*, the oxidation may be brought about by heating the pig iron with oxide of iron or

[1] Sterry Hunt.

substances containing it, such as red hematite, hammer-scale, best tap-cinder, etc. The iron oxide gives up a portion of its oxygen to the impurities, and they are removed into the slag.

Practically the same thing occurs when an air blast is employed. The oxygen first forms oxide of iron, which is subsequently decomposed by the silicon, etc., present.

Upon these principles, all processes for the conversion of pig iron into mild steel and wrought iron depend. The essential difference lies in the fact that while *mild steel* is obtained at the end of the process in a *molten* state, and is run into moulds, *wrought iron* is obtained in an unfused, *spongy* condition (owing to the lower temperature that prevails), the particles of iron being subsequently welded together.

The order of affinity for oxygen is as follows: silicon, carbon, manganese, phosphorus, iron, sulphur. If it were possible to make the oxidizing influences act uniformly through the metal, the impurities would be removed in this order. In the Bessemer process, where air is blown into the molten pig iron, something closely approaching this takes place (see p. 147), but in other methods overlapping occurs.

Sulphur is not removed by oxidation, but is taken up by the slags, apparently by liquation, as sulphide.

Processes for producing wrought iron by subjecting pig iron to a blast of air in hearths, are known as *finery* processes; those in which it is treated with oxides of iron in a reverberatory furnace, as *puddling*.

Refining.—All kinds of pig are not equally suitable for conversion into malleable iron. Only some 80 per cent. of the phosphorus present, and 40 per cent. of the sulphur, are removed in finery and puddling processes. The presence of too much silicon also gives trouble owing to the fluidity of the metal, and consequent difficulty of working, as well as the great loss in weight which occurs.

The pasty stage through which white pig iron, free from manganese, passes prior to fusion, permits of easier mixture with the iron oxides and oxidizing slags in the furnace, which effect its purification; and in consequence of its composition less loss occurs. It is therefore preferred to grey iron on this account, but, as its sulphur contents are usually higher, this advantage is often more than counterbalanced. Unless the iron is smelted from pure ores, it is now usual to produce grey pig in the blast furnace, and either treat this directly or subject it to a refining process previous to its actual treatment for malleable iron.

Refining is a process formerly generally employed for the

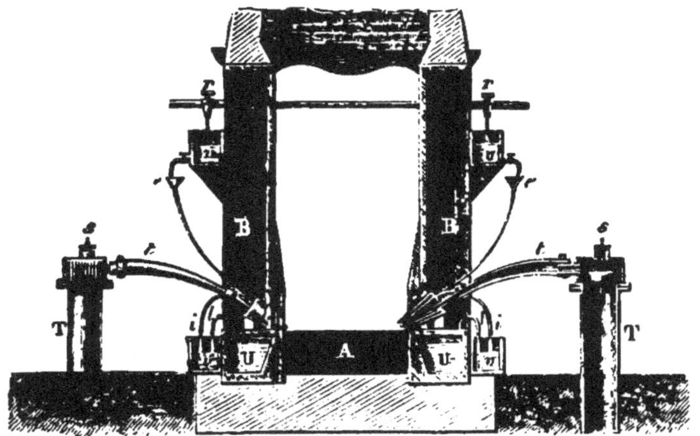

FIG. 43.

conversion of grey into white iron previous to puddling or blooming.

The refinery or running-out fire is shown in Fig. 43. It consists of a rectangular hearth 4 feet square and 18 inches deep, formed on three sides of water-cooled cast-iron blocks, U U. The front is a cast iron plate in which the tap-hole is situated. Four columns, B B, situated at the corners of the hearth, carry the brickwork stack—some 16 to 18 feet high—on girders. The bottom is made of sandstone blocks.

The hearth is enclosed by iron plates attached to the columns. Those at the back are hinged, while the front plate is attached to the end of a lever and counterpoised, so as to be easily raised and lowered. The hearth is provided with a number of water-cooled tuyeres, usually five or six.

These are inclined at an angle of about 30°, and placed on both sides in such manner that the tuyeres are not opposite to each other. In this way the blast is uniformly distributed over the hearth. In front is an iron mould for the reception of the metal, and beyond this a pit for the slag, which runs off the surface when the mould is full. Owing to its lower melting-point, it remains fluid longer than the iron.

The blast is employed at a pressure of about 2½ lbs. per square inch.

The hearth, being hot from a previous operation, is partly filled with coke, and the charge of about two tons of pig iron and scrap introduced, in layers alternating with coke, through the folding doors at the back. Some hammer-scale (Fe_3O_4) is often added. The blast is turned on, and in about a couple of hours the charge is melted down. More coke is added if necessary, and the blast continued for from half to three-quarters of an hour, during which period bubbles of carbon monoxide are seen escaping from the metal and burning on top. The metal, when deemed refined, is tapped out and cooled rapidly, water being often thrown on the surface for this purpose. The plate of metal is from 1 to 3 inches thick.

Owing to the large amount of air supplied, the iron is subjected to oxidizing influences during the whole period of the operation, and, when fluid, the downward direction of the tuyeres causes the blast to play continually on its surface. The oxide of iron thus formed, and that added as hammer-scale, attacks and oxidizes the silicon, carbon, and phosphorus in the metal.

The removal of silicon from the metal in refining is most marked, pig iron containing 5 per cent. being reduced to from 0·5 to 0·7. Carbon is seldom reduced more than 1 per cent., while the removal of phosphorus is very variable. In some cases it is scarcely affected. The rapid cooling ensures the retention of the remaining carbon in the combined form. A white, close, dense metal results known as *plate metal*, or *refined iron*.

The slag consists mainly of basic silicate of iron. In this process sulphur is not removed.

Removal of Sulphur from Pig Iron.—The purification of pig iron from sulphur has received much attention. Manganese and sodium carbonate are both employed for this purpose. In each case, a sulphide not decomposed by iron is formed.

With sodium carbonate, silicon is also largely removed, and some carbon, with separation of metallic sodium. The molten pig is run into a receiver containing the carbonate.

Scheerer proposed the use of calcium chloride and salt as a desulphurizer, by adding it in the puddling process. **Saniter's** desulphurizing process consists of running the molten pig into a receiver containing calcium chloride and lime. Fluor spar may also be added.

NOTE.—The "washing" process is a method of refining pig iron for steel-making.

Finery Processes.—Welsh Finery, Walloon Process, Swedish-Lancashire Hearth.

In these processes, the pig iron is converted into malleable iron in open hearths, *in contact with the fuel.* In consequence, only charcoal can be employed, coke and coal not being sufficiently free from sulphur.

The Swedish-Lancashire hearth is a small rectangular finery made of cast-iron plates. The top is arched over, and communicates with a chamber in the flue in which the pig iron is heated prior to being placed on the hearth. The hearth has one tuyere, nearly horizontal, supplied with blast heated to about 120° C., by circulating through iron pipes placed in the flue.

The hearth is filled with charcoal, and the charge of about 2 cwts. of white or mottled iron drawn from the flue, the blast turned on, and the charge melted down. The atmosphere is highly oxidizing, and as the drops of metal sluggishly pass before the tuyere oxidation occurs.

The metal collects at the bottom, and slightly hardens. The cake is broken up by the workmen, and held before the tuyere, to be remelted and further oxidized. When the metal gets stiff and infusible, it is raised to the top, fresh charcoal added, the temperature raised, and the whole remelted. As it drops down before the blast into the bath of highly basic slag at the bottom, its fining is completed, and the pasty mass is collected into a ball, withdrawn from the furnace, and consolidated by hammering, the retained slag being thus expelled.

The Walloon process is similar. These finery processes are wasteful. A loss occurs of from 15 to 20 per cent. on the pig iron employed.

They are still retained in Norway and Sweden, and were formerly in vogue in South Wales, for the production of tin bars for tin-plate manufacture, but are now superseded, tin bars of better quality being made from open-hearth steel at a much reduced cost.

Puddling.—This process—the most important of all methods of making malleable iron—was introduced by Cort, in 1784, as a method of employing coal for fining iron. Up to this period, the sulphur in coal and coke had prevented their use in fineries.

The employment of reverberatory furnaces, in which the fuel is burnt out of contact with the iron, completely overcame the difficulty; the sulphur, being burnt to SO_2, has no effect on the iron.

Fig. 44 shows a section of a puddling furnace. It is a reverberatory furnace, the grate area of which is large in proportion to the hearth space (1 : $1\frac{1}{2}$ or 2). The bottom

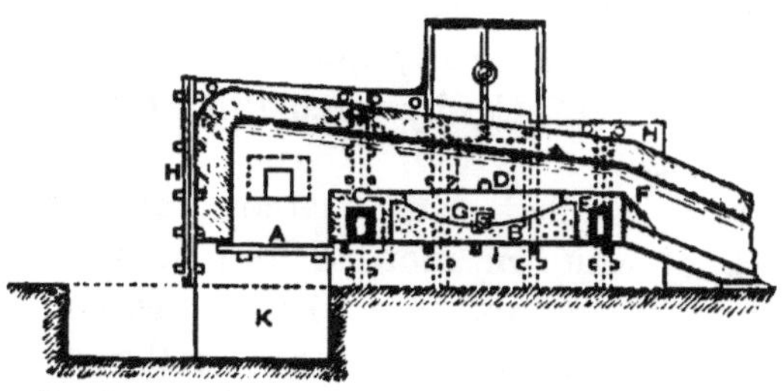

FIG. 44.—A, fire-place; B, bed; C, fire bridge; D, working door; E, flue bridge; F, flue; G, tap-hole for slag; H, plating of furnace.

and sides of the hearth consist of cast-iron plates suitably supported, and backed with fire-brick. They are protected with some material rich in oxide of iron, and kept cool by the circulation of air under and round them.

Formerly brick beds and sand bottoms were employed.

The working door, in front, is only opened to introduce and withdraw the charge. It slides between guides, and is attached to the end of a lever, and is counterpoised. The working of the charge is effected through an opening in the bottom of the door (the stopper-hole), through which the bars for this purpose are introduced. In front of the door is a shelf or fore-plate. The interior is lined with fire-brick, and the furnace is supported externally by iron plates and tie-rods. A damper in the flue or at the top of the stack is employed to regulate the draught.

The fire- and flue-bridges are usually hollow, the circulation of air keeping them cool. These, as also the sides of the hearth, are sometimes kept cool by the circulation of water. Under the fore-plate is a tap-hole, for the removal of slag, which is generally run out every second heat.

The iron plates forming the bottom and sides of the hearth are protected by about 3 or 4 inches of "fettling." Bull-dog, calcined pottery mine, and best tap-cinder[1] are employed for this purpose. They are broken to about the size of macadam, and spread on the bottom, the spaces being filled up by similar material ground fine and moistened. The fire-brick lining above the side plates projects slightly, so as to retain the fettling. Puddler's mine (soft red hematite) and blue billy (ferric oxide produced in burning pyrites for making sulphuric acid) are used for making the bed even. All these substances soften at the furnace temperature, and play an important part in the operation.

Bull-dog is a mixture of ferric oxide (Fe_2O_3) and silica obtained by roasting tap (puddling furnace) cinder. This consists of a very basic silicate of ferrous oxide. On roasting, the FeO takes up oxygen, and is converted into Fe_2O_3, which has very little affinity for silica, and separates from it.

It is fairly refractory, except in a reducing atmosphere, when the Fe_2O_3 is reduced to FeO, which at once combines with the silica.

With a new bottom, it is usual to introduce a quantity of light wrought-iron scrap (bustling), gradually raise it to a welding heat, and work it into a ball. The oxide produced is spread over the bed. This is repeated every shift of twelve

[1] It is the slag from reheating furnaces.

hours, but the bed is repaired after every "heat" by introducing fresh fettling wherever necessary.

The fuel used is a free-burning coal. In furnaces using steam-jet injectors, small and inferior coal may be employed, and a great saving effected. The large grate space is necessary to attain the high temperature required in the furnace. The depth of the fire is about 10 inches. When anthracite is employed, it is less, while the grate area is increased, and the size of the flue diminished.

The charge generally consists of from 3 to 5 cwts. of pig iron, with the addition of hammer-scale (Fe_3O_4), etc., as may be required.

The process may be divided into four stages: (1) Melting-down stage. The pig is placed near the fire-bridge, the fire made up, and the damper opened. The manner in which the iron melts, and the temperature, depend on whether grey or white iron is being treated. Grey iron requires a higher temperature than white, and at once becomes very fluid. White iron passes through a pasty state prior to perfect fusion. During this stage, silicon and phosphorus are principally removed by oxidation.

(2) The boiling stage. When all is melted, the damper is lowered to somewhat reduce the temperature, and the metal, which slightly stiffens, is thoroughly mixed with the oxides of iron (hammer-scale, mill cinder, etc.) added and produced during the melting-down stage. The oxides of iron in the slag and lining of the furnace rapidly oxidize the remaining silicon and also the carbon in the metal. The temperature of the metal rises, and the whole surface is agitated by the escape of the carbon monoxide generated. Each bubble as it emerges bursts into flame (puddler's candles). During this period, the puddler is continuously employed in rabbling (stirring up) the charge, thoroughly incorporating it with the oxides of iron in the cinder. The boiling gradually subsides, and the metal becomes stiffer and quieter. The carbon has been reduced below 1 per cent., and the third stage supervenes. It is usual in some districts to remove the greater part of the slag at this stage by skimming it off the surface.

(3) The fining stage. During this stage, the remainder of the carbon and the manganese are removed, together with

some phosphorus. The movements of the metal by CO, owing to its pasty state, are sluggish. The metal is raised and broken up from time to time, and the damper is raised to thoroughly liquefy the slag. The fluid cinder sinks, and bright points gradually extending over the surface, caused by the burning of the iron, show themselves, and announce that the metal has "come to nature."

(4) Balling stage. The pasty, spongy mass of malleable iron, now at a full welding heat, is then made into balls, weighing some 70 lbs., by pushing the spongy iron together, and rolling it over the bed of the furnace. As these are made, they are rolled up to the fire-bridge end, and the damper is lowered. The atmosphere of the furnace thereby becoming smoky and reducing, oxidation and waste of the iron is to a large extent prevented. These balls are then removed separately on an iron bogey to the hammer or squeezer, and *shingled*—that is, the particles of malleable iron welded together and the slag squeezed out.

The whole operation takes ordinarily about $1\frac{1}{2}$ hour, divided as follows: 30 to 35 minutes, running down; 10 to 15, boiling; 10 to 20, fining; 20 to 30, balling up and shingling; but may be longer or shorter according to the purity of the metal treated.

The loss varies from 7 to 20 per cent., according to purity. The greatest loss occurs with siliceous pig, such as is treated in Scotch forges.

The process described above is that followed with grey forge pig, and is technically known as **pig boiling**, the principal decarburizing agent being the oxides of iron in the lining and slag, the air only affecting the metal in the running down and balling stages. Some of the chemical changes which take place may be thus expressed—

$$Fe_3O_4 + Si = 2FeO,SiO_2 + Fe$$
$$2Fe_2O_3 + 3Si = 3SiO_2 + 2Fe_2$$
$$Fe_2O_3 + C + SiO_2 = 2FeO,SiO_2 + CO$$
$$2FeO, SiO_2 + C = FeO,SiO_2 + CO + Fe$$
$$2FeO,SiO_2 + O = Fe_2O_3 + SiO_2$$

The silica formed is fluxed off by oxide of iron, of which there is always large excess, and the carbon is removed as carbon monoxide.

Manganese is oxidized to MnO, which replaces FeO in the slags, rendering them very fluid. The phosphorus is removed by oxidation, and passes into the slag as phosphate of iron. It is probable that to some extent it liquates into the slag as phosphide of iron, and is subsequently oxidized.

Dry Puddling, as it is called, as now conducted, differs but little from the pig-boiling process, save that white or refined iron is the metal operated on, and in consequence the action is less vigorous, and the amount of slag less. The temperature is lower throughout until the balling stage is reached, the metal never becoming perfectly fluid, and being continuously rabbled. The decarburization is mainly effected by the air passing through the furnace. Formerly it was conducted on a sand bottom. Owing to the nature of the iron employed, there is, of course, less loss.

Tap Cinder, as the slag from puddling is called, consists of basic silicate of iron, with smaller quantities of lime, alumina, oxide of manganese, and phosphoric acid. Sulphur, probably as iron or manganese sulphide, is also present. It is a black mass, with a dull granular fracture. It may be represented by the formula $2FeO,SiO_2$. In the puddling process it acts as a carrier of oxygen to the impurities in the pig, the ferrous oxide it contains being oxidized and subsequently reduced. It contains from 40 to 60 per cent. of iron, and is tapped from the furnace into rectangular iron waggons. An inferior class of pig iron, known as *cinder pig*, is made from it by smelting in the blast furnace.

The sulphur in pig iron is not removed by oxidation either in puddling or finery methods. A considerable portion, however, does pass into the slag, probably by liquation. Its removal is facilitated by any cause which tends to prolong the operation or to render the slag fluid. For this reason, manganese in iron, by prolonging the fining stage and the effect of the oxide formed in giving greater fluidity to the slag, promotes the elimination of this element. It appears to pass out during the whole operation. Various "physics" are employed. **Schaffhautl's** powder is a mixture of oxide of manganese, salt, and clay. **Scheerer's** consists of calcium chloride and salt, soda-ash, etc.

Improvements in Puddling.—Besides the introduction of steam-jet injectors for forcing the fire, various contrivances to diminish labour and save fuel have been introduced. Mechanical rabbles, which traverse the hearth, imitating more or less perfectly the motion given to the rabble by the puddler, have been introduced. In all cases, however, the charge must be balled up by hand. Mechanical furnaces, in which the working of the ball is effected by the revolution of the chamber, have also been employed. The most successful of these is Dank's furnace, for a description of which some larger manual must be consulted. In Pernot's furnace, the *hearth only* revolves in a plane slightly inclined to the horizontal.

Puddling furnaces heated by gas, and provided with regenerators on the Siemens principle, have also been employed.

The waste heat from puddling furnaces is generally used for raising steam.

Shingling and Rolling.—The "balls" taken from the puddling furnace consist of a sponge of malleable iron saturated with slag. The consolidation and welding together of the particles, and the expulsion of the slag, are technically known as *shingling*. The freedom of the wrought iron from slag will depend upon the efficiency with which the operation is conducted. This process takes the form of hammering or squeezing.

The *crocodile squeezer* is shown in Fig. 45. It consists of

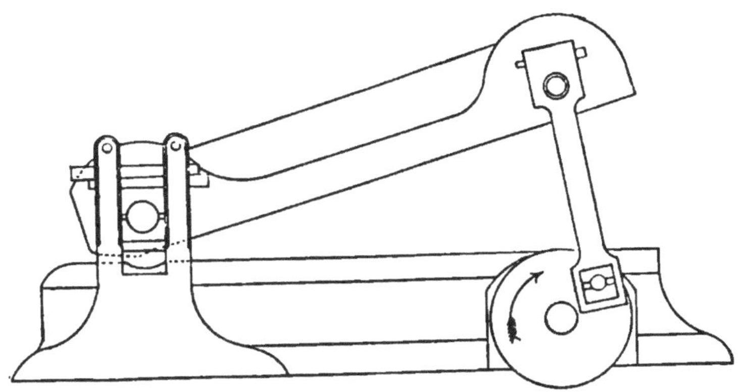

FIG. 45.

two jaws, the lower of which is fixed and forms the anvil, while the upper opens and closes upon it, actuated by the revolution of the crank. The ball is placed between the open jaws, and turned over and moved towards the back part of the jaw as the mass gets reduced in bulk and consolidated by the expulsion of slag, which flows out over the sides of the anvil. Many other forms of squeezer are also employed.

The *helve* or *shingling hammer* is shown in Fig. 46. The head, weighing about 8 to 10 tons, is raised (by the cams on the wheel revolving in front), from 15 to 20 inches, and allowed to fall on the ball placed on the anvil-block. The number of blows is from 60 to 100 per minute. In belly-helves, the cams act on the lever at a point midway between the head and the fulcrum.

The objection to the helve is that the weight of the blow is the same at the beginning of the operation, when the ball is tender, as when it has become more solid.

Steam-hammers are now almost universally employed.

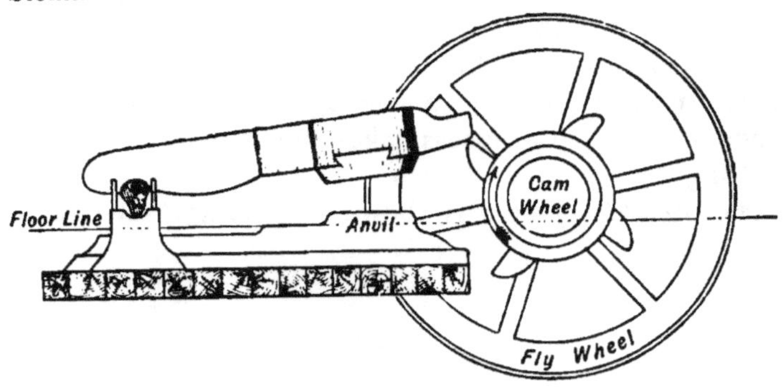

FIG. 46.

They consist of an inverted upright steam cylinder, to the piston rod of which the head or "tup" is attached. This slides between vertical guides on the standards which support the cylinder. The admission of steam to the cylinder is controlled by a handle which works the valves through a system of levers. In double-acting hammers, steam is admitted below the piston to raise the head, and also above the piston to force it down and increase the weight of the blow. In single-acting hammers, steam is only admitted below the piston to raise the tup, which then falls by its own weight. The larger hammers for forging purposes are of the former type.

The legs of the shingler are encased in iron guards, and the face protected by an iron mask, to protect him from slag, which flies about in all directions. The ball is placed on the anvil, and receives at first a few light blows. This is effected by admitting a little steam under the piston as the head falls, thus forming a cushion and diminishing the force of the blow. The weight of the blow is gradually increased, the ball being turned at each stroke, until it has been hammered into a rectangular "bloom," or billet, and the slag thoroughly expelled. The bloom is still hot enough to be rolled out into bars, and is dragged over the iron plates of the floor to the "puddle rolls," or forge train.

These consist, as shown in Fig. 47, of two pairs of iron rolls, 15 to 18 inches in diameter, mounted in suitable housings, the lower one being driven directly from a steam-

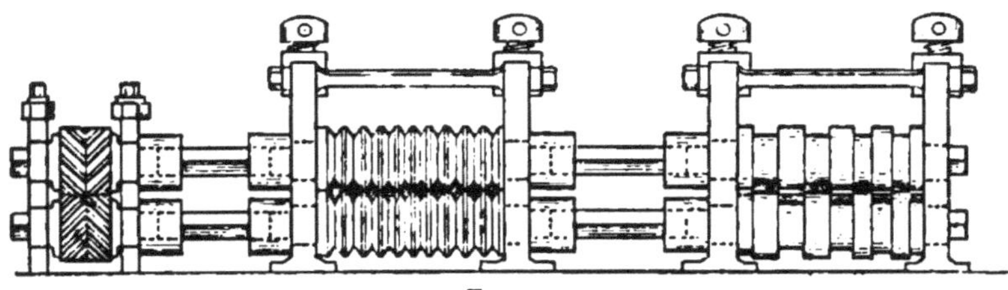

FIG. 47.

engine. One pair of rolls, known as the *roughing* or *cogging* rolls, has a series of gothic, or V-grooves, of diminishing size; and the other, known as the *finishing* rolls, a series of rectangular grooves.

The surface of the grooves in the roughing roll is notched or chisel-cut, to enable them to grip the bloom and drag it through. It is first pushed endwise into the widest of the grooves in the roughing rolls. As it leaves the rolls on the other side, it is seized and passed back by resting it on the top of the upper roll, which, by its revolution, carries it forward. It is then passed through the next groove, and this is repeated until it has been reduced to the required size. It afterwards makes one or more passes through the rectangular grooves of the finishing roll, and is thus reduced to a flat bar, which is dragged aside and allowed to cool. It then constitutes "puddle bar," on the weight of which the puddler is paid. The fracture is bright and crystalline, or granular. The puddle rolls make about seventy revolutions per minute. The surface of the rolls and the bearings are kept cool by jets of water.

Puddled bar is never homogeneous, and includes particles of slag not expelled during shingling.

Crown or Merchant iron, is produced by cutting the puddle bar into suitable lengths and arranging them in a "bundle," "pile," or "faggot," which, if large, is tied by iron wire. The size of the pile depends on the size of the bars, etc., to be produced. It is raised to a full welding heat in the *reheating* or *mill furnace*, which somewhat resembles a puddling furnace, but is without a flue-bridge.

Reheating furnaces working with gas, and provided with regenerators, are now commonly employed.

When fully hot, the piles are withdrawn, and the bars welded together under the hammer or in a blooming-mill, the pile being afterwards reduced to a size suitable for rolling.

It then passes to the *mill train*, which consists, as before, of two sets of rolls—roughing and finishing. The pile is cogged down to the required extent in the roughing rolls, and then passes to the finishing rolls to be converted into rounds, squares, angles, or any other form (section) required. The finishing train has chilled iron rolls, and the grooves are turned with great accuracy. Sometimes, instead of at once finishing, the billet, after rolling down, is cut up, and piled, reheated, and again rolled. It then forms No. 3, or best iron, while, if this is again piled and reheated, best best iron results.

The oxide of iron formed in the reheating combines with the sand of which the bed is made and forms a slag, which flows out of the flue towards which the bed inclines. It is known as *flue cinder* and *mill-furnace* slag. It consists of ferrous silicate, with a large excess of oxide of iron, and has a lustrous crystalline fracture.

Light work is guided into the rolls by various devices, and is consequently known as *guide iron*.

In rolling plates, plain rolls are employed. The billet is passed in one direction until the required width has been obtained, and then turned at right angles and rolled down to the desired thickness.

The distance between the rolls is regulated by setting-down screws, which act on the top bearing of the upper roll; and in rolling plates, the distance between them is diminished at each pass, both ends being set down by the same amount.

The weight of the upper roll is counterbalanced. The roughing rolls are of grain iron, but the finishing rolls are chilled on the surface. Heavy plate mills are either provided with reversing gear, or are driven by reversing engines, so as to obviate passing the work back over the top roll.

Thin sheets are rolled by doubling and passing the compound sheet through the rolls. Sometimes as many as sixteen thicknesses are being rolled at one time (see Tin Plate).

For light work, to save the time of passing it from the back to the front, and the consequent cooling down which takes place, three-high rolls are employed. The middle one is driven from the engine, and the others geared with it. The work having passed through the lower pair is returned through the upper. Hoops, for example, are thus made.

Three-high mills for heavy work are provided with rising and falling tables, which receive the work as it leaves the rolls, and raise or lower it as required.

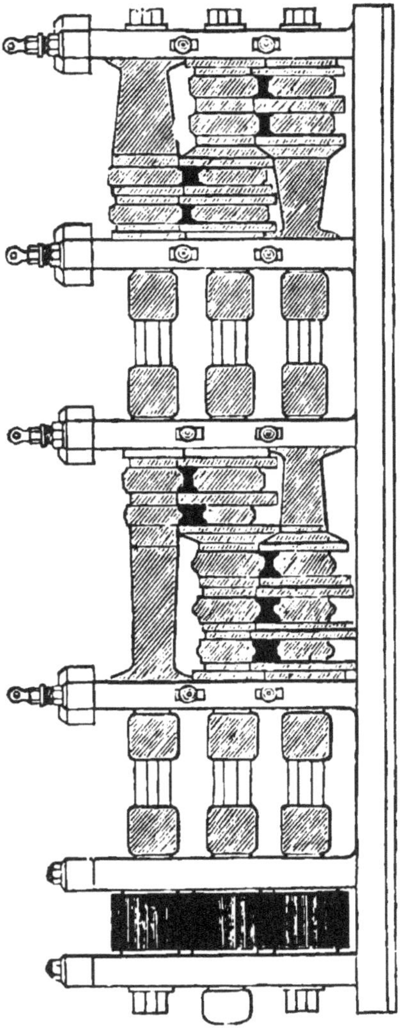

FIG. 48.—Three-high Rail Mill.

Rolls for rolling finished iron vary from 8 inches to 38 inches in diameter.

The rolling of malleable iron by welding together and

elongating the particles of iron develops a fibrous structure, which is more pronounced the greater the number of times it is piled and reheated. This treatment also renders the metal more uniform in character.

COMPOSITION OF MALLEABLE IRON.

Carbon	0·1 to	0·3
Silicon	traces ,,	0·1
Phosphorus	0·04 ,,	0·2
Sulphur	0·02 ,,	0·15
Manganese	traces ,,	0·25
Iron	99·1 ,,	99·8

Burnt Iron.—When iron is exposed at a very high temperature to an oxidizing atmosphere, it loses its malleability and is known as *burnt* iron. Probably this is due to the formation of a suboxide of iron in the metal.

Brands of Merchant Iron.— (Crown), common iron, or merchant bar (puddle bar, once piled and reheated). Best, twice piled and reheated. Best best, three times piled and re-heated. Treble best, four times piled and reheated.

CHAPTER XI.

STEEL.

THE designation of steel was formerly confined to those varieties of iron which could be hardened by heating to redness and plunging in cold water.

The introduction of the Bessemer process marked a new era. The metal produced by this process lacked the fibrous character associated with wrought iron, and partook more or less of the character of steel. Those varieties possessing more than 0·3 per cent. of carbon sensibly harden when treated in the same manner as steel, but with less carbon this is not the case. Other processes producing similar soft metal sprang up, and the term *steel* has come to include a great variety of material having widely different properties. Some are softer even than wrought iron, and cannot be hardened.

Since the hardening property is dependent on the amount of carbon it contains, a classification based on the percentage of that element is the most convenient, steel containing less than 0·5 per cent. being classed as *mild* steel. Steel proper contains from 0·5 to 1·5 or 1·7 per cent. of carbon. The different nature of these metals may be shown by the use of such titles as Bessemer, Siemens's or open-hearth steel. Some of these contain as little as 0·13 per cent. of carbon, less than is often present in wrought iron. They differ from that metal in being devoid of fibre, more homogeneous, and, unlike it, *are obtained in a state of fusion*, and cast in ingots. The term *ingot iron* would be more applicable than *steel*.

Steel.—The fracture of steel becomes finer the larger the proportion of carbon present, but is affected by such treatment as hammering cold. Steel of hard temper,[1] breaks with a bright, uniform, bluish grey, finely granular fracture. After hardening, the colour is somewhat whiter.

It is very malleable, but requires working more carefully and at a lower temperature than wrought iron. Steel containing less than 1·25 per cent. of carbon can be welded. A lower temperature must be employed than for malleable iron, or the steel will be burnt. To render the surfaces clean at the lower heat, borax mixed with about one-tenth of its weight of sal ammoniac is employed to dissolve the scale.

The specific gravity varies from 7·624 to 7·813, in the unhardened state, to 7·55 to 7·75 in the hardened condition, showing that expansion occurs.

The melting-point varies with the proportion of carbon. The softest melts a little below 1600° C. The hardest at about 1400°.

The *tenacity* varies from 22 tons in mild steel to upwards of 70 tons in steel of hard temper. Its *elasticity* exceeds that of wrought iron, while its *ductility* is equal to the best qualities of that substance. The mild varieties suffer an elongation and diminution in area, when subjected to a stretching force, greater than wrought iron. The elongation of the harder varieties is much less, but the elastic limit is high.

[1] The term "temper" applies only to the proportion of carbon present.

Hardening and Tempering.—The extent to which hardening occurs depends on the proportion of carbon in the metal, and the rate and manner of cooling.

Thus, quenching in mercury or other good conductor of heat produces greater hardness and brittleness than quenching in water, while quenching in oil (oil hardening) produces a degree of hardness, without brittleness (owing to the slower cooling action of the oil), whereby the tenacity of the steel is increased. Gun tubes are treated in this way.

Hardened steel may be rendered soft by heating for a prolonged period at a high temperature, and allowing it to cool down very slowly. This is called annealing.

When steel, rendered brittle by hardening, is heated at temperatures below redness, the hardness is partially removed, and it recovers to some extent its elasticity; the higher the temperature attained, the softer and tougher will the hardened metal become. This operation is known as *letting down* or *tempering*. If the surface of the hard steel be polished, on gradually heating in air, it becomes first a pale straw colour, and afterwards dark straw, golden yellow, brown, brown with purple spots, purple, violet, and blue, as the heating proceeds. These tints serve to mark the degree of heat attained, and, in tempering tools and cutting instruments, indicate to the workman the point at which they should be cooled off. The colours are probably due to thin films of oxide formed on the surface. The hardness which they indicate depends on the nature of the steel. Below is a table giving the temperatures indicated, and some articles let down to the respective tints.

220° C. light straw: lancets, razors, and surgical instruments.
230° C. dark straw: surgical instruments, razors.
245° C. full yellow: penknives, wood tools, taps, dies.
255° C. brown: cold chisels, hatchets, etc.
265° C. brown with purple spots: axes, plane-irons, pocket-knives.
275° C. purple: table-knives, large shears.
295° C. violet: swords, watch-springs, augers.
320° C. full blue: hand and pit saws, etc.

The cause of these changes in hardness is the manner in which the carbon exists in the metal. In the unhardened state, the carbon is in the form of carbide of iron (Fe_3C). On heating to redness this carbide is decomposed, and in the hardened metal the carbon is present in some other form, known as "hardening carbon," diffused through the metal. The degrees of hardness depends on the ratio between the "carbide" and "hardening" carbon. In tempering, some of the hardening carbon changes to carbide, the amount depending on the degree of heat attained.

VARIOUS QUALITIES OF STEEL.

Description.	Percentage of carbon.	Character and uses.
Mild steel	0·1 to 0·25 (0·2 ,, 0·4 Mn)	Soft malleable metal for rivets and plates.
	0·3 ,, 0·4	Harder and stronger for rails, forgings, etc.
	0·4 ,, 0·5	For tyres and castings.
	0·5 ,, 0·6	For hard wire, for guide ropes, springs, etc.
Die temper	0·75	Is tough, capable of resisting great pressure, is very easily welded. Used for stamping and pressing dies, welding steel for axes, plane-irons, etc.
Sett temper	0·825	Is hard, tough, strong, capable of resisting sudden and great shocks, blows, etc. Used for cold setts, minting dies, and smiths' tools; easily welded.
Chisel temper	1·0	Is easily forged; hard even when let down low, and sufficiently tough to withstand blows; is weldable. Used for cold chisels, miners' drills, large punches, etc.
Punch temper	1·125	Is a hard, fine-grained metal. Takes and maintains a good cutting edge; is more difficult to work, but welds with great care; used for circular cutters, taps, rimers, large turning tools and drills, screwing dies, etc.
Turning-tool temper	1·25	Is unweldable, and must be carefully treated in forging, hardening, and tempering; is generally useful for turning, planing, and slotting tools, drills, small cutters, taps, saw-files, etc.
Razor temper	1·5 and upwards.	This and the last variety are unsuited for any purpose where sudden variation in pressure, etc., occurs. Can only be dealt with by a skilful workman, and if at all over-heated, is spoilt. Used for razors, surgical instruments, small tools, etc.

The harder varieties of steel occupy an intermediate position, between malleable iron, on the one hand, and cast iron on the other. The comparison ends with the carbon contents of the metal, as in all steels, save special kinds to be noted hereafter, the other elements present in cast iron, are found only in minute quantities. They exist in appreciable amounts in the metal obtained from all processes in which steel is made direct from cast iron.

Steel-Making.—The methods of producing steel may be classed as follows :—

1. *Direct methods*—
   (*a*) From iron ores.—Catalan and analagous processes.
   (*b*) From cast iron.—Puddled steel.
2. *Indirect methods*—
   (*a*) By the carburization of malleable iron in an unfused state.—Cementation and case-hardening processes.
   (*b*) By carburization of molten malleable iron.
      (1) Fusion of bar iron with carbon in crucibles.—Cast crucible steel and Wootz processes.
      (2) The carburization of molten malleable iron obtained by complete or partial decarburization of pig iron. Bessemer and open-hearth processes.

Steel in the Catalan Forge.—Excellent steel of middle temper can be made in open hearths of this type, by giving the tuyere less inclination, so that the blast does not play so directly on the accumulating mass of metal, and removing the slag more frequently than in making malleable iron.

In making *steel* in these hearths, less small ore is added, so that the slag is less basic, and less blast employed. By these means the reduction is somewhat prolonged, affording opportunity for carburization by decomposition of carbon monoxide by the spongy iron, while the direction of the tuyere and the removal of the slag prevent decarburization by the oxide of iron it contains, and by the air. The presence of manganese in the ore is also favourable to the production of steel. Its oxide gives greater fluidity to the slag, and is less energetic as a decarburizing agent.

Puddled Steel.—Steel can be made in the puddling furnace by arresting the process before decarburization is complete, sufficient carbon being left to constitute steel. White pig irons containing manganese and free from sulphur are best adapted for the purpose.

Cementation Process.—This is by far the most important method of producing steel for cutting-instruments, *i.e.* steels of hard temper. As previously noted, when iron is heated in contact with carbon, carbon monoxide, or compounds of carbon and hydrogen (hydrocarbons), to a high temperature, carbon is taken up by the iron. This is the basis of the process by which all the best qualities of steel for cutlery, springs, etc., are produced. The superiority of the method is due to the fact that practically pure iron is employed for the

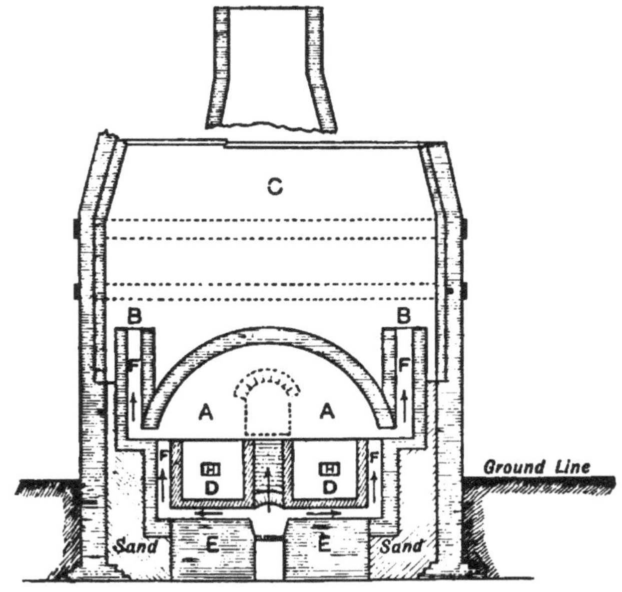

FIG. 49.

purpose. Swedish bar iron, made in the Swedish Lancashire hearth under charcoal, from charcoal pig, is the material usually operated on, so that such steels practically consist solely of iron and carbon. The bars employed are about 10 feet long, 3 inches wide, and $\frac{5}{8}$ inch thick. Hammered bars are preferred. Basic steel bars are also sometimes employed.

The converting furnace is shown in Fig. 49. The furnace proper consists of a rectangular arched chamber, A, of fire-brick. This communicates by the chimneys B B B, three on each side, with the hovel C about 40 feet high, which serves as a chimney and diminishes the loss of heat by radiation. It gives the furnace the appearance of an ordinary glass furnace. A narrow fireplace, 12 to 15 inches in width, runs down the middle, with a firing door at either end. On each side of the fireplace is a *trough* or *pot*, D, for the reception of the bars of iron. They are made of firestones, open at the top, and rest on the benches E E on brick bearers, which divide the space below the pots into a number of flues, F. These are continued up the sides and ends of the pot; the space above the fire is similarly divided, so that the boxes may be heated as uniformly as possible.

The pots are from 10 to 15 feet long, $3\frac{1}{2}$ to 4 wide and deep. In the end of each is a small opening, H, known as the tap-hole. This is opposite a similar opening in the outer wall, and through it the *tap* or trial bars are withdrawn, from the appearance of the fracture of which, the progress of the operation is judged. A manhole is provided for the purposes of charging and discharging the pots, and is closely bricked up during the conversion.

The pots are charged by first spreading a layer of charcoal nubs (about as large as peas or beans) over the bottom. On this a layer of bars, about half an inch apart, is placed. A second layer of charcoal, followed by bars, is then put in, and so on, until the pots are full, finishing off with charcoal. The charge is covered with "wheelswarf."

This is the refuse from under the grindstones, and consists of particles of oxidized (rusted) iron and sand. At the high temperature of the furnace, this frits and forms a rough glass, which hermetically seals the pots and excludes air.

The manhole is bricked up and carefully luted, as also is the space round the trial-bars. A coal fire is then made and the temperature gradually raised. In about 24 hours the pots are at a dull red heat, and in about 50 hours or more, the bright red or yellow heat (1100 to 1200° C.) required for conversion is attained. This is steadily maintained for a period of from 4 to 8 or 10 days, depending on the degree of carburization required. For springs, saws, etc., 4 or 5 days suffice. For shear steel, 5 or 6 days; double shear, 7 to 8 days; and tool steel, 10 days or more. The progress is judged by the appearance of the fracture of the trial bars, a crystalline layer of steel of greater or less depth is formed, enclosing a "sap" of unaltered iron, but there is no sharp line of demarcation. When judged complete, the fires are allowed to burn out, and the furnace to cool very gradually. This occupies about a week, and the pots are then discharged. The bars present a blistered or warty appearance, and a laminated structure, and are hence known as *blister steel.*

These blisters are evidently formed by the efforts of gas to escape from the interior of the bar while in a pasty state. The gas is formed by the action of the carbon on particles of slag, containing oxide of iron, enclosed in the iron.

The bars are brittle, and are sorted out by breaking with a hand hammer on a block, and examining the fracture. No. 1, "spring temper," shows a comparatively thin skin of steel enveloping unaltered iron. In No. 4, "double shear heat," the proportions of steel and iron are about equal. In No. 6, "melting heat," the "sap" has disappeared, and the conversion has extended through the bar. The carburization is probably due to the decomposition, by the iron, of carbon monoxide produced from the oxygen in the small amount of air retained in the pot and in the pores of the carbon (see page 107). As before noted, iron at redness is easily permeated by gases, and thus carbon is carried into the interior of the bar. The amount of carbon taken up varies with the time and temperature up to 1·5 per cent.

Blister Steel is brittle, largely crystalline, and lacks homogeneity. For most purposes it is either tilted or melted.

Shear Steel.—The blister is cut into lengths, faggoted, reheated, and welded, and then either drawn out under the hammer or rolled, in a manner resembling the treatment of iron. By this means greater uniformity of composition is obtained. The metal sometimes undergoes a second piling and reheating.[1] By this treatment the percentage of carbon is very slightly reduced by oxidation, and only the milder

[1] After once piling it is known as *shear steel*, and after a second treatment, as *double shear steel.*

tempers with less than 1·125 per cent. of carbon can be satisfactorily welded. The pile is frequently coated with clay wash and borax, to protect it from oxidation and facilitate welding. Tilted steel has lost the laminated appearance of blister steel and is more uniform in character.

Cast Crucible Steel.—Steel produced as above must necessarily be far from homogeneous. In 1740, Huntsman introduced the practice of melting down the blister steel in crucibles, pouring it into ingot moulds, and working the ingots into bars, etc. The fusion ensures uniformity of character and composition, hence the term *homogeneous*

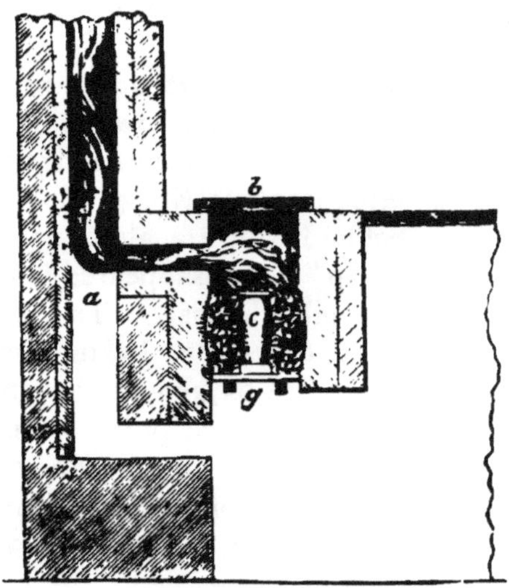

FIG. 50.

metal or steel applied to it. Its commoner designation is *crucible cast steel.*

The steel-melting holes or fires (see Fig. 50) are simple wind furnaces of oval section, lined with ganister. They are placed below the floor level for convenience of handling the pots. Each fire has a separate flue, which is continued down behind the furnace, and opens into the ash-pit. The draught is regulated by the insertion or removal of a brick in this opening. The crucibles, which are dried on shelves round the melting-house, are from 16 to 19 inches high and 6 to 8 inches in diameter at the mouth. Each fire takes two pots. Before placing them in the furnaces they are annealed in a stove or oven, mouth downwards, for some hours, gradually attaining a dull red heat. The blister steel is cropped up into small pieces, and the charge is introduced into the heated crucibles by means of an iron funnel. The pots generally last about three melts, the weight of the charge being less each time. Thus a first charge of 50 lbs. will be followed by charges of 45 and 40 lbs. respectively for the second and third.

The charge having been introduced, the lid is put on, the fire is made up with hard, free-burning coke, and the furnace closed. This first fire burns off in about forty-five minutes, and is followed by a second and third firings. The amount of fuel added in the third fire is judged from the amount of metal remaining unmelted, to ascertain which, the workman pokes an iron bar into the pots and gives directions accordingly, in order that all the crucibles may be ready at one time. The crucibles are lifted from the fires, for teeming, by grasping them round the belly with tongs having bent jaws, which encircle the pots. The first melt occupies from 4 to 5 hours.

Small ingots are run from a single pot. For larger ones the pots are *doubled*, that is, the contents of two pots are transferred to one before teeming, while for still larger ones the metal from the crucibles is transferred to a ladle similar to that described (see p. 150), or arrangements must be made to keep up a constant stream of metal into the mould.

The ingot moulds are of cast iron, and made in two parts. While casting they are held together by an iron ring. The moulds are warmed and reeked, that is, coated with lamp-black, by smoking them with the flame of burning tar. Sometimes a wash of clay is applied. This treatment prevents the ingots from sticking. In pouring, the hot stream of metal should not touch the sides.

The pots, if in good condition, are returned, after detaching clinker, etc., to the fires, ready for the next charge. If allowed to cool, they cannot be reheated without cracking. In melting blister steel, it is usual to add a small quantity of black oxide of manganese, which is partly reduced, and manganese passes into the metal. The slag is removed before teeming by moving a knob of slag (mop), attached to an iron bar, over the surface of the metal, by which means the slag is cooled, collects on the mop, and is removed.

Direct Cast Crucible Steel.—In casting large ingots of crucible steel, bar iron or puddled steel is employed instead of blister steel, charcoal, spiegel, and ferro-manganese being added to carburize the metal to the desired degree. Ingots 40 tons in weight have been cast.

Fig. 51 shows a regenerative crucible furnace for steel melting.

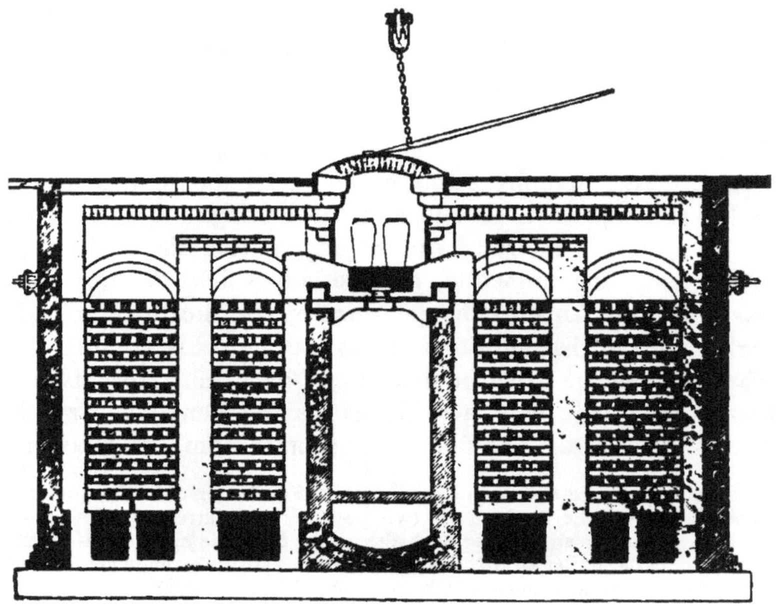

Fig. 51.—Regenerative Crucible Furnace.

They take from 8 to 24 pots in a double line. The roof is in several sections, which can be removed as required for charging or teeming purposes. Some furnaces of this type are provided with a movable bottom, which can be elevated, by a hydraulic ram, to the floor level, with all the pots standing on it.

Plumbago pots, having a larger capacity than those described, are also employed. They are more durable than ordinary white or black pots (*i.e.* clay, or a mixture of clay and coke-dust), and with care will bear reheating after cooling. They last from 9 to 11 melts.

Honeycombing.—Steels, especially those of mild temper (below 0·5 per cent. of carbon) are liable to boil up in the mould after teeming. This is due to the disengagement of dissolved gases, mainly H, N, and CO, which are given off as the metal cools. The bubbles of gas cause the metal to be honeycombed and vesicular. With a view to prevent boiling up, a stopper, which fits loosely, is put on the top of the metal, and a little sand thrown on and round it to keep it down; or sand is thrown on the top of the metal and an iron cover is put on and held down by wedges passing through eyes on the top of the mould.

The upper part is most affected, the gases from the lower part rising, in consequence of the metal at the bottom remaining fluid for a longer time. The presence of the stoppers prevents the top from cooling so rapidly.

Piping.—Steel of harder temper (above 0·7 per cent. of carbon) settles down in the mould, forming a funnel-shaped cavity or pipe. The ingot is *topped* by breaking off the unsound part before working. Both these evils may be largely mitigated by careful melting and teeming at the proper temperature.

Dead-melting.—If not heated a sufficient length of time, the metal is not "killed," and will teem "fiery"—throwing off sparks—and be honey-combed. If "dead-melted," this does not occur, but if left too long in the fires it will teem "dead," and be weak and brittle.

Case-hardening.—The surfaces of wrought iron and mild steel articles subject to wear, are often superficially hardened by packing them in iron boxes, with parings of horns and hoofs, leather scrap, bone-dust, and charcoal, and heating to full redness. The depth of the hardening depends on the length of time they are heated. Small articles are case-hardened by heating them to redness and sprinkling or rubbing them in powdered yellow prussiate of potash (potassium ferro-cyanide). The carburization is effected by the cyanogen (CN) compounds.

Production of Steel from Pig Iron without previous conversion into malleable iron.—These processes involve the removal from the pig of the silicon, sulphur, and phosphorus, and the reduction of the amount of carbon to the quantity required to convert the metal into steel. It is found more satisfactory, however, to completely remove carbon as well, and recarburize by the addition of carbon in some form or other, generally as spiegeleisen or ferro-manganese; but gas carbon and other substances are also employed (Darby process).

Bessemer Process.—In the Bessemer process the impurities are burnt out of the pig by blowing air through the molten metal. Referring to p. 121, it will be seen that all the impurities, except sulphur, will practically be oxidized before the iron, so that by stopping the blast at the right moment, and adding a quantity of spiegeleisen or other carbon-bearing material, sufficient carbon may be introduced to produce steel of the desired temper.

The process is generally conducted in a vessel or converter of the form shown in Fig. 52. It consists of a boiler-plate casing $\frac{3}{4}$ to 1 inch thick, carried on a cast-iron ring, provided with trunnion arms, upon which it is carried in bearings on standards or other supports. Upon one of the trunnions is keyed a toothed wheel, which gears with a rack (Fig. 53) attached to a hydraulic ram, by the movement of which the converter can be rotated on its bearings through 180° to 300°. The other trunnion is hollow, and connects by the pipe P (Fig. 52) with the blast-box B at the bottom of the converter. This is a compartment into which the blast is led through the hollow trunnion, and forced through the metal by means of clay tuyeres T passing through the upper or guard-plate of the blast-box and the lining of the vessel. The vessel is lined with about 9 to 12 inches of ganister on the sides, and 18 to 20 inches on the bottom, introduced as described, p. 34. The tuyeres are slightly conical in form, and are about 22 inches long. They are made of fire-clay, and contain from 10 to 12 $\frac{3}{8}$-inch holes, running in the direction of the length, by which the air passes from the blast-box to the vessel. They pass up through holes in the guard-plate, against which they are pressed by suitable stops, and are embedded in the ganister lining the bottom. They only stand out slightly from the surface.

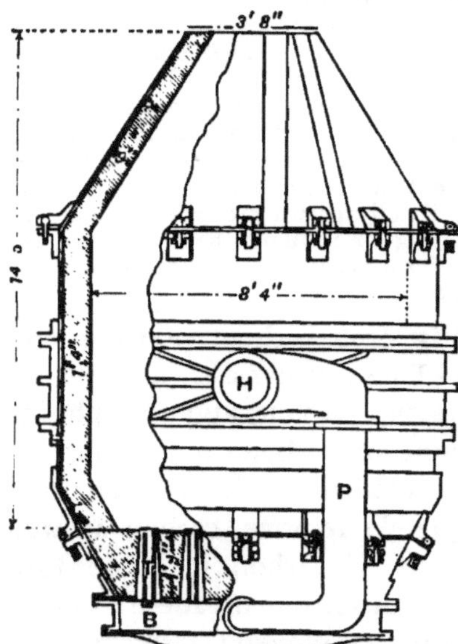

Fig. 52.—Bessemer Converter, as used in Basic Process.

If a tuyere proves faulty in work it can be removed and replaced, by taking off the bottom plate, knocking it out, and pushing up a new one in its place, a little slurry of ganister being run round from the inside to make the joint secure. After drying and heating the converter is again ready.

Converters with detachable duplicated bottoms are now commonly employed, so that little delay is occasioned by the removal of a worn-out or faulty one, and the substitution of a newly prepared bottom. The vessel itself is also made in sections, as shown, and duplicate parts are kept in stock.

Method of conducting the Process.—The pig iron to be treated is melted in cupolas, or is taken direct from the blast furnace, after mixing to ensure uniformity. The converter, previously heated, is turned on its side, and the metal run in. The full charge lies below the level of the tuyeres when in this position. The blast, at a pressure of from 20 to 25 lbs., is then turned on, and afterwards the vessel is rotated into a vertical position. The metal now flows over the bottom, and the air passes up through it, the high pressure preventing it running into the blast-box. At first only a short, yellowish-red flame is seen at the mouth of the converter, accompanied by sparks. During this period the temperature rapidly rises. The silicon is being rapidly oxidized to silica (SiO_2), which, combining with oxides of iron and manganese, forms silicates. The flame gradually becomes larger and more luminous, and is accompanied by showers of brilliant sparks, consisting of slag and particles of iron. This corresponds to the "boiling stage" of the puddling process, and is known as the *boil.* The violence of the disturbance of the metal is due to the rapid oxidation of the carbon with the production of carbon monoxide, which escapes. During this part of the process, the pressure of the blast is reduced. The luminosity and volume of the flame gradually diminish, and in the third or "fining" stage, during which the remainder of the carbon and manganese are being removed, it fades to a pale amethyst tint, and is nearly transparent. There are also fewer showers of sparks. In from fifteen to twenty minutes from the commencement of the blow, the flame suddenly shortens or "drops." This marks the almost complete removal of the carbon, and if the blast is further continued, great loss from oxidation takes place, and the quality of the metal is rendered much inferior. The vessel is accordingly turned down and the blast shut off. A weighed quantity of spiegeleisen, previously melted in a cupola, is added to the metal as it lies in the converter, from a ladle. This addition is attended by a violent outburst of flame and considerable agitation of the metal. The spiegel imparts to the iron the requisite amount of carbon to produce steel of the desired temper, and also sufficient manganese to restore the malleability, which, as before noted, is always lost when malleable iron in a molten state is subject to oxidizing influences. For steel of very low temper, ferro-manganese is employed, in order to introduce the necessary amount of manganese without adding too much carbon. This is added solid. After standing a few minutes to allow the slag and metal to separate, the converter is turned down and the steel run from its mouth into the ladle. Enough slag to cover the metal and keep it hot is also allowed to flow into the ladle. The converter is then turned completely over, and the slag allowed to run out. All the movements of the vessel, as also the blast, are regulated by a workman situated on an elevated platform at some distance from the converter, the

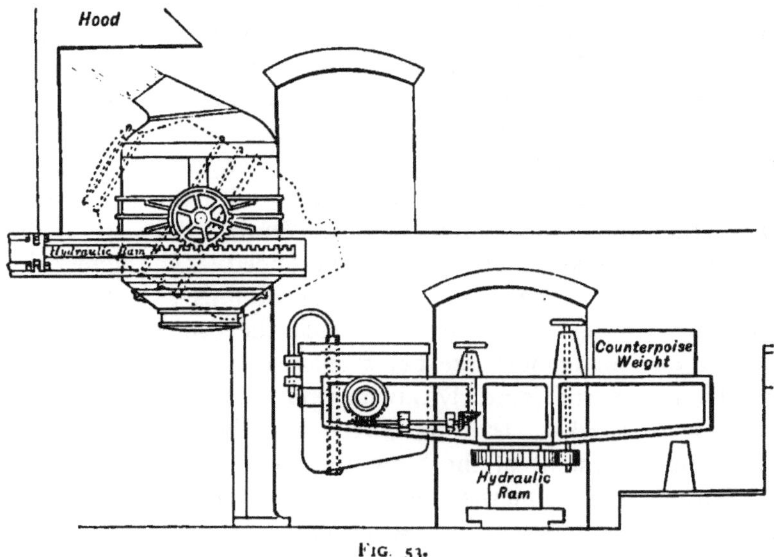

FIG. 53.

progress of the operation being judged from the appearance of the flame.

The ladle to which the metal is transferred is mounted as shown (Fig. 53) on a hydraulic crane, in the centre of the casting pit, which is circular. The converters are situated at the side of the pit. The ladle can be raised and lowered, can be made to travel round the pit, to and from the centre, and also turned over to empty slag. It is lined with ganister, and heated by a fire made in it before receiving the metal. The teeming is effected from the bottom, through a hole closed by a fire-clay stopper, which is raised and lowered by an iron rod protected with fire-clay tubes and connected with a suitable lever.

The moulds are of cast iron, open top and bottom, and more or less tapering. They are arranged round the side of the casting pit, standing on an iron plate. The usual practice is to fill each mould separately from the top, but sometimes the moulds are arranged in groups, round a central one somewhat taller than the rest. The bottom of this is connected by a system of fire-clay tubes opening upwards, with the bottoms of the others. The metal is run into the central or feeding ingot and is conveyed to the others by the clay passages. It gradually rises in the moulds until they are all filled. Sounder ingots are said to be obtained in this way. In all cases they are stoppered down with sand and a plate, as previously described.

Chemical Changes in the Bessemer Process.—The oxidation of the impurities in the pig is effected by oxide of iron, formed by the air blown in, so that the chemical reactions are similar to those taking place in puddling. Being in a fluid state, and thoroughly agitated by the passage of the air, silicon, being most oxidizable, is much more largely attacked than other impurities in the first stage of the process, and is reduced to about 0·5 per cent. In the subsequent stages it is reduced to about 0·02 or 0·03 per cent. During the boil the carbon is reduced to below 1 per cent., and in the fining stage to below 0·1 per cent. Manganese is attacked from the beginning and throughout the process, the oxide formed combining with the silica and forming silicate, which passes into the slag. The phosphorus, as shown by its absence from the slags, is unattacked, and the steel consequently contains a higher proportion than the original pig iron, since a loss of some 10 per cent. occurs on the weight of pig employed. This is due, as already noted on p. 36, to the siliceous nature of the lining. The iron employed must therefore be free from phosphorus. Sulphur also is not removed.

The blast is used cold, and yet the temperature gradually rises as the process proceeds. This increase of heat is due to the oxidation going on, principally of the silicon in the iron. The amount of silicon present is less than the carbon, but in burning, a solid substance (SiO_2), which remains behind in the converter, is produced, and all the heat generated is communicated to the contents of the vessel (save such as is carried away by the nitrogen of the air blown in). The combustion of the carbon generates gaseous bodies which, escaping, carry away much of the heat. Manganese, like silicon, yields a solid product of combustion, and accounts, when present, for some of the heat.

During the blow the iron is oxidized and becomes "burnt," and is rendered brittle and unworkable. The manganese in the spiegel added combines with the oxygen, forming manganous oxide (MnO), and passes into the slag. A slight excess is always employed to ensure the complete removal of the oxygen. This, with the carbon contained in the spiegel, enters the steel. Bessemer and open-hearth steels always contain manganese. The amount should not exceed 0·5 per cent.

The metal employed is grey pig iron, and should contain about 2 to 2½ per cent. of silicon, and be free from sulphur and phosphorus. Iron smelted from pure ores, such as red hematite and magnetite, and known as *Bessemer pig* is employed. With the rapid, continuous working practised in America—a fresh quantity of pig being run in immediately after teeming the previous charge, that no loss of heat occurs—pig containing not more than 1 per cent. of silicon is satisfactorily treated. Excess of silicon increases the amount of loss.

The process described above is commonly known as the *acid* process, from the siliceous nature of the ganister lining. The slag is a basic silicate of iron and manganese. As already shown, iron containing phosphorus cannot be treated under these circumstances. By substituting a lining of basic material, phosphorus as well as other impurities may be removed.

The Basic Bessemer Process is conducted in a vessel similar to that already described, but generally with a straight neck, so that the metal can be poured from either side, and the converter can be completely rotated by worm and wheel gearing, actuated by hydraulic engines attached to the standards.

The converters are made in sections, which can be readily secured together by pins and cotters (see Fig. 52), so that if the lining of any portion gives way another similar part can be substituted without delay, an overhead travelling crane which commands the converters, and hydraulic tables under each converter, being provided for raising and lowering the parts.

The lining employed consists of calcined dolomite or magnesite (see p. 37), and is about 14 to 16 inches thick on the sides, and 24 inches on the bottom. Loose tuyeres are sometimes employed, but generally they are formed by ramming the lining material round steel spikes, which are withdrawn when the bottom is rammed up, and thus form free passages for the air. The process differs somewhat from the ordinary "acid" process. Before running in the iron, a quantity of lime, equal to about 15 per cent. of the charge, is introduced, with a little coke, into the hot converter, and blown hot. The charge is then run in, in the usual manner, and the blow proceeds as before up to the point at which the flame drops. Instead of stopping the process here, the blast is continued for some two or three minutes longer to eliminate the phosphorus.

The vessel is then turned down, and a sample taken with a spoon, hammered out, cooled, and broken. From the fracture and malleability, the workman judges how long the blow must be continued to complete the elimination of phosphorus. A crystalline fracture indicates that the phosphorus is not completely removed, and the vessel is turned up, and the blowing continued until the metal is dephosphorized. A second sampling may be necessary.[1]

The slag is then run off, to prevent precipitation of phosphorus into the metal by reduction from the slag when the carbon is added. Spiegel and ferro are then added in the usual manner, and the charge transferred to the ladle, and thence to the moulds. In some cases, where hard metal is required, the carburization is effected by grey pig iron free from phosphorus, added in a molten state to the metal in the ladle, ferro-manganese being afterwards added.

The oxidation of impurities during the blow, up to the dropping of the flame, proceeds as in the acid process, but, owing to the nature of the lining and the basic character of the slag, some phosphorus is also removed. In the after-blow the residue of the phosphorus is oxidized, and, combining with lime, forms calcium phosphate, and passes into the slag. This frequently contains as much as 30 per cent. of phosphates of lime and magnesia, together with 8 to 10 per cent. silica, 10 per cent. oxide of iron, sulphur, and some oxide of manganese. It amounts to about 20 per cent. of the charge, and on account of the phosphates present is ground up and used as manure.

If the pig iron treated contains much silicon, the charge becomes too hot, and the corrosion of the lining is increased. Since, under ordinary conditions, the presence of silicon is essential to provide heat by its oxidation, some substitute is necessary in basic Bessemer pig. This is found in the phosphorus, and as much as 1·5 to 3 per cent. of that element is often present. Some of the slag is returned to the blast furnace to increase the phosphorus in the pig. It lowers the melting-point of the metal until decarburization is complete, and thus less heat is requisite in the earlier stages. By its oxidation in the after-blow, it produces the high temperature necessary to maintain the iron in the fluid state. As with silicon, the product of burning is a solid, and remains in the vessel.

---

[1] The time during which the blowing is continued after decarburization is known as the "*after-blow*." During its continuance, red-brown smoke issues from the converter.

About 1 per cent. of silicon is necessary in the iron, or the blow will be too cold and slow. The presence of from 1 to 2 per cent. of manganese is also advantageous.

The removal of the phosphorus depends on the basicity of the slag, and hence the addition of lime in the converter. This also diminishes the wear on the lining. The loss amounts to about 15 per cent.

The charge of a converter ranges from 5 to 15 tons, and the operation lasts from 15 to 25 minutes, according to weight and circumstances.

Open-hearth Processes.—Under this heading are included processes conducted in *regenerative gas furnaces* of the Siemens type (see Fig. 54), the bed of which may be composed of silica sand (acid), or of magnesite, dolomite, or chromite (basic).

Siemens's Regenerative Furnace is shown in Figs. 54 and

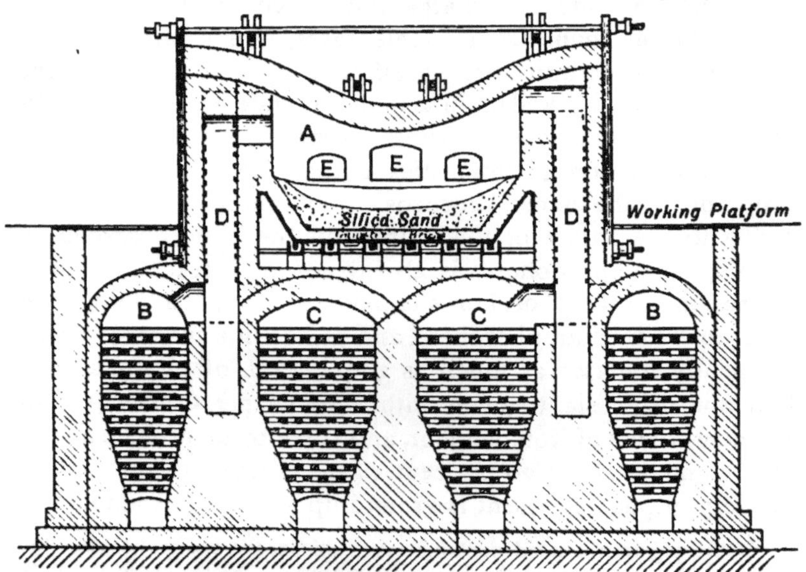

FIG. 54.—Siemens's Regenerative Furnace. Longitudinal Section.

55. The furnace is a double-ended, reverberatory, gas-fired furnace. The furnace chamber, A, communicates at either end with the chambers B B, C C, by means of the ports and flues D D. The chambers are filled with chequer brickwork, built up of bricks 2 inches square. The chequers are alternately heated by the passage of the hot gases from the furnace descending through them on their way to the chimney-stack, and the heat retained is subsequently given up to the cold air and gas passing

upwards through them on their way to the furnace. The chambers are worked in pairs, the gas and air being heated in separate chambers. One pair of chambers is being heated up while air and gas are passing through the other pair. The smaller chambers, B, are the gas chequers, and C the air chequers; E E E are the working doors; F is the chimney-flue; G the gas-supply culvert; R R the valves for reversing

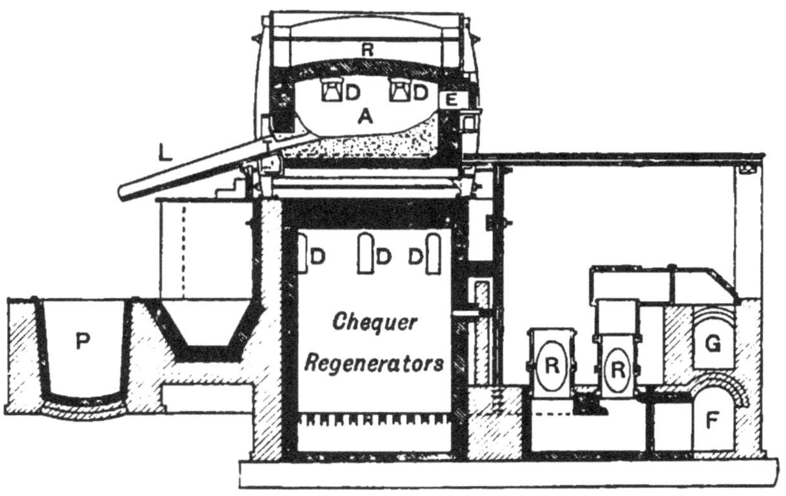

FIG. 55.—Cross-section of Furnace.

the direction of the gas and air; L the launder or spout for conveying the metal into the ladle; P is the casting pit. The direction of the air and gas are reversed every half-hour. In this way the chequers are kept at a high temperature, and the gas and air coming to the furnace, develop a much higher temperature than if supplied cold.

The Siemens Process.—This process is analagous to the "pig boiling" puddling process, the decarburization of the metal being effected by *pure* oxidized iron ores added to the fused metal in the bath of the furnace. On this account it is sometimes described as the "pig and ore" process.

The pig iron, to the extent of 5 to 40 tons, is introduced on the bed of the furnace and melted. After fusion, additions of red hematite, roasted pottery mine, or other pure oxidized ores are made from time to time, which effect the oxidation and removal of the silicon, carbon, and manganese in the pig in the same manner as in puddling.

At the high temperature attainable in these furnaces, however, the metal remains molten even after decarburization is complete, and its conversion into steel is effected by the addition of spiegel and ferro, as in the Bessemer and Basic Bessemer processes. The time occupied is, however, much longer, extending sometimes to 10 or 14 hours with large charges. This permits of more perfect control over the composition of the steel produced, as samples can be taken from time to time, and the character and carbon contents of the metal rapidly determined. When the carbon has been reduced below 0·1 per cent., spiegel is added, the tap-hole is broken open and the metal run into the ladle. Some ferro-manganese, broken into small pieces, is generally added in the ladle as the metal flows out, to replace that lost by oxidation in the furnace, and to make up the amount necessary to restore the malleability and carburize the iron.

In the decarburizing stage, the metal boils violently, and is thus brought into contact with the oxidizing slags and the atmosphere of the furnace, but becomes quiet towards the end of the operation. The addition of the spiegel causes it to again become lively, and the metal is tapped on the boil.

The yield is some 2 or 3 per cent. in excess of the pig iron charged, owing to the reduction of the ore added to decarburize it. A cutting oxidizing flame is employed in the earlier stages.

Siemens-Martin Process.—In this process the percentage of carbon to be removed from the metal is diminished, by melting the pig iron with scrap wrought iron or steel introduced into the furnace at the same time, or previously heated and charged into the bath of molten pig iron. Scrap to the extent of 8 or 10 times the weight of pig is frequently employed. The charge, after fusion, contains less than 1 per cent. of carbon. The amount of scrap added depends on the greyness of the pig. No ore is added, and the decarburization is effected by the oxide formed on the scrap during melting, and the atmosphere of the furnace, which is oxidizing. The bath is sampled from time to time, and, when the carbon has been sufficiently reduced, spiegel and ferro-manganese are added as before. The loss amounts to about 7 or 8 per cent. of the metal charged.

A combination of the two processes is commonly used in this country, pig iron, scrap, and ore forming the furnace charge. It affords a convenient method to utilize scrap.

Basic Open-hearth Processes.—In furnaces with sand bottoms the pig iron employed must be of Bessemer quality, but with basic bottoms phosphoretted pig can be treated. As in the basic Bessemer process, lime is charged in the furnace, and samples are taken from time to time and tested.

As the phosphorus is not required as a heat-producer, the less there is present the better. Pig containing about 1·5 to 2 per cent. is satisfactorily treated, but the presence of manganese up to 2 or 3 per cent. is also desirable, as it prolongs the fining stage and permits of the elimination of the phosphorus without undue oxidation of the iron. In dephosphorizing, it is sometimes necessary to make small additions of ferro-manganese and pig to prevent this. The metal obtained by any of these processes is dealt with as in the Bessemer process.

Casting.—Formerly the casting-pits were rectangular, and the ladle, mounted on a carriage, travelled on rails over the top of ingot moulds.

Circular or semicircular casting-pits with hydraulic central cranes are being introduced.

Attempts have been made to combine the rapidity of the Bessemer process with the certainty of the results obtained in the open-hearth processes.

A combination of the two processes is followed by blowing the metal in a converter till the carbon is sufficiently reduced, and then teeming it into a heated Siemens's furnace, and completing the decarburization in the ordinary manner.

Hollow rabbles introduced into the molten metal on the hearth, by which air or steam can be blown through it, are in use to a limited extent. Clay-covered iron tubes are employed. At Ruhort, 3 such tubes, each containing 3 holes, are employed. The blast is continued for from 10 to 20 minutes, and the temperature rises higher than in the ordinary open-hearth process.

The steel produced by the Bessemer, Siemens, and analagous processes is generally of a mild character, containing less than 0·5 per cent. of carbon, and is employed for rails (0·3 to 0·4 per cent. of carbon); boiler-, bridge-, and ship-plates, 0·2 to 0·25 per cent. of carbon; rivet-iron, 0·1 to 0·15 per cent. of carbon; armour-plates, guns, and other purposes where metal of high ductility, elasticity, uniformity, and strength are required, and also for castings.

The honeycombing previously noted in connection with crucible steel is more marked. Stoppering down is resorted to, and in some cases pressure is applied.

In Whitworth's fluid compressed steel, the ingot, after running in a mould of special construction, is placed on the table of a hydraulic press and subjected to a pressure of from 6 to 20 tons to the square inch. A contraction of 1½ inch per foot takes place, and a sounder ingot results.

At Krupps', the pressure of liquid CO_2 is employed for the same purpose. The ingot-moulds are provided with a gas-tight cover, through which a narrow pipe connected with a metallic reservoir of liquid CO_2, and provided with a stop-cock, passes. On warming the reservoir in water, great pressure is exerted.

At the Edgar Thompson works, compression by high-pressure steam has been tried, steam at a pressure of from 100 to 200 lbs. per square inch being employed.

In the production of sound steel castings, various physics are resorted to. The introduction of from 0·2 to 0·3 per cent. of silicon in mild tempers, and from 0·3 to 0·4 per cent. in harder tempers, tends to increase the solidity of the casting. It is introduced as ferro-silicon or ferro-silicon manganese—alloys of silicon with iron and manganese, but containing also carbon. Aluminium is also employed for the same purpose.

Small wheels are cast on revolving tables, making some 50 or 60 revolutions per minute: the metal is run into the mould at the centre. The rim is thereby rendered denser.

It has also been proposed by Mr. Allen to stir the metal in the ladle with a revolving paddle, to disengage the gas prior to casting.

Treatment of Ingots.—The ingot moulds, after the solidification of the metal, are lifted by cranes situated at the side of the casting-pit, and the ingots allowed to cool; or, in the newer works, removed immediately to "soaking-pits," in which they are kept hot till required for rolling.

These soaking-pits consist of a series of vertical chambers of fire-brick below the ground-level, arranged in a double line, each capable of holding an ingot, and covered with a tile and commanded by cranes. The ingots are removed to them immediately they have solidified.

The interior of the ingot when stripped is much too hot to permit of it being rolled at once, and the excess of heat gradually soaks out and distributes itself uniformly through the mass. They can be kept hot for some time, and removed for rolling as required. Little heat is lost, and reheating of the ingots is avoided. Oxidation is prevented by the gases exuding from the metal, which are of a reducing character (p. 146).

The difficulty of keeping up a supply of ingots to keep the pits hot has led to the use of "soaking-furnaces," the several cells or pits of the furnaces communicating with each other, a fireplace or gas-producer being provided at one end of the system.

The rolling of mild steel is effected in a manner similar to that followed for malleable iron.

Use of Spiegel and Ferro-manganese.—In carburizing Bessemer or open-hearth steel, the richness in manganese of the alloy used is mainly determined by the amount of carbon desired in the resulting steel. If a steel very low in carbon is required, an alloy (ferro-manganese) containing much manganese is employed, to introduce the needful amount of that element, without at the same time adding an excess of carbon.[1] For higher carbon steel, spiegel and ferro containing less manganese are employed. Steels containing a higher percentage of carbon than 0·5 may be made, as in the Darby process, by carburizing with gas carbon, anthracite, etc. The molten metal is run into a ladle containing the carburizing material, which it dissolves. An undue proportion of manganese is thus avoided.

CHAPTER XII.

COPPER.

Physical and Chemical Properties.—This metal possesses a fine red colour, and is characterized, when pure, by extreme toughness. Its hardness is slightly under 3. It is more malleable but less ductile than iron. In the cast state, its tenacity is only about 9 to 12 tons, but after rolling this is increased from 15 to 18 tons, and by wire-drawing to 30 tons. Its modulus of elasticity as wire is 17,000,000, that of iron wire being 25,300,000. Its specific gravity is 8·6, but by rolling, etc., is increased to 8·8. Its melting-point is about 1050° C. Pure copper is an excellent conductor of heat and electricity, but the presence of minute quantities of impurity greatly impair this quality. The metal

[1] The amount of *carbon* present in spiegel and ferro shows no *great* variation.

is unaltered in dry or moist air free from CO_2 and acid vapours. A green coating of basic salts forms under these circumstances.

On heating in air, a series of coloured oxide films are formed, and at a red heat a black scale of oxide, which detaches itself when the metal is suddenly cooled. The outside of this scale consists of black cupric oxide (CuO), but the inner layers principally consist of red cuprous oxide (Cu_2O). Cupric oxide is reduced to cuprous oxide when fused with copper. Copper dissolves cuprous oxide when molten, and is rendered dry and brittle. "Dry" copper breaks with a dull, brick-red fracture. In practice the metal is toughened by covering with anthracite, and stirring it with birch poles. Hence the term "poling." The reducing gases from the wood, in conjunction with the anthracite, reduce the oxide, and the metal assumes its normal tough condition.

If the copper is not chemically pure, it is possible to continue the poling too long, and the metal again becomes dry and unmalleable. By leaving a little of the oxide unreduced, the harmful effects of the impurities are neutralized to some extent. It is described as "underpoled," "tough cake," and "overpoled," according to its condition. Pure electrotype copper cannot be overpoled or burnt. *Underpoled* copper contracts greatly on solidifying, and a furrow is formed down the middle of the ingot. *Tough cake* copper casts with a nearly flat surface, while, if *overpoled*, the metal rises in the mould, and a ridge is formed.

Very small quantities of lead, arsenic, sulphur, antimony, and bismuth, seriously impair the malleability, ductility, and tenacity of copper. Tin, nickel, cobalt, and iron, are also often present in commercial copper. They render the metal lighter in colour and somewhat harder, but do not lower its tenacity.

Copper has greater affinity for sulphur and less affinity for oxygen, than iron has. Two sulphides are known. *Cuprous sulphide*, Cu_2S, is the "white metal" of the copper smelter, and is produced by heating copper and sulphur together. It occurs naturally in various copper ores.

Cupric sulphide (CuS) is precipitated when a soluble sulphide is added to a solution containing copper.

The sulphides of copper are *not* reduced by iron or carbon. When heated in air, sulphur burns off as SO_2, and a mixture of oxides and sulphate, in proportions varying with the conditions, results.

Sulphate of copper is soluble in water, and is decomposed when strongly heated. It requires a higher temperature to decompose it than sulphate of iron. When sulphide of copper is heated with oxide or sulphate, the sulphur and oxygen pass off as SO_2 and the metal is reduced.

$$Cu_2S + 2Cu_2O = SO_2 + 3Cu_2$$
$$Cu_2S + CuSO_4 = 3Cu + 2SO_2$$

Copper and phosphorus combine readily, forming phosphide of copper.

Bronze includes all alloys of copper and tin.

The effect of tin in whitening copper is greater than that of any other metal. The alloys have a lower melting-point, and cast sounder than copper. The toughness, tenacity, and other properties vary with the composition of the alloy. (See Alloys, p. 268.)

Brass includes all alloys of copper and zinc. The effect of zinc in whitening is much less than tin, and hence wider range of colour is possible.

The malleability and tenacity of certain of these alloys is little inferior to copper, *e.g.* Dutch metal is beaten into thin leaves in imitation of gold, and brass, for wire and plate, has a tenacity of 8 or 9 tons cast, which, after rolling and wire-drawing, varies from 20 to 26 tons. (See Alloys, p. 267.)

Ores of Copper.

(1) **Native copper** often occurs with copper ores, sometimes in masses, as in the Lake Superior district, but more often in arborescent and reticulated forms. In the Calumet, Hecla, and other mines, about 2 per cent. of native copper, mainly in small grains, is distributed through the rock.

This is extracted by dressing processes, and smelted and

refined at one operation. The copper Barilla of Chili, was a deposit of grains of copper, oxidized on the surface. Native copper is usually very pure.

Cuprite, *red oxide of copper, cuprous oxide* (Cu_2O) occurs, crystallized and massive, in Thuringia, Chessy, near Lyons, Cornwall, Siberia, United States, Cuba, Australia, etc. It contains 88·8 per cent. of copper when pure.

Tenorite, *black oxide of copper* (CuO) occurs extensively in Chili and Australia. It is generally very impure.

Green Malachite is a hydrated carbonate of emerald green colour, of the composition $CuCO_3,CuH_2O_2$. It is often beautifully variegated, and is used for ornamental purposes. It occurs in Siberia, Australia, and United States. It contains 58 per cent. of copper.

Blue Malachite, *Azurite* or *Chessylite* ($2CuCO_3,CuH_2O_2$), is of a deep blue colour, and generally occurs with green malachite. An extensive deposit formerly existed at Chessy in France. It contains about 55 per cent. of copper.

Chrysocolla and Dioptase are hydrated silicates of copper. The former is blue and the latter green in colour. They contain about 30 per cent. of copper.

Redruthite, *copper glance* (Cu_2S), occurs native in Cornwall and elsewhere. It has a semi-metallic white appearance, and is readily scratched with a knife. It contains 80 per cent. of copper.

Erubescite, *Bornite, horseflesh ore* ($3Cu_2S,Fe_2S_3$), occurs extensively in South Africa, Australia, and Norway. It consists of copper and iron sulphides, containing up to 62 per cent. of copper. Its colour varies from copper red to pinchbeck brown, with a blue tarnish.

Copper Pyrites, *yellow copper ore* (Cu_2S,Fe_2S_3), is distinguished by its golden yellow metallic appearance. It is softer than iron pyrites, and can be scratched with a knife. When pure it contains 34·6 per cent. of copper, 30·5 iron, 34·9 sulphur. Usually, however, it is mixed with a large excess of iron pyrites (FeS_2), and does not contain more than 12 per cent. of copper, and often less. It is the principal English ore of copper, and occurs abundantly in Cornwall and Devonshire, also in Siberia, in Sweden at Fahlun, in the Hartz, and various localities in the United States.

Peacock Copper Ore is a variegated copper pyrites, but is usually richer in copper.

Grey Copper Ore, *tetrahedrite, Fahl ore*, consists of sulphantimonides and sulpharsenides of copper and iron. It often contains silver, mercury, and sometimes gold. The amount of copper varies up to 38·6 per cent. It occurs extensively in the Hartz Mountains, at Kremnitz in Hungary, at Frieberg in Saxony, Kapnuik in Transylvania, and in Chili. It is worked for copper and also for silver.

Atacamite is a natural oxychloride, occurring extensively in Atacama in Chili, in Australia, and elsewhere. It is a deep green in colour.

Cupreous Iron Pyrites.—Besides the above, much copper is extracted from the cinders from the burning of cupreous iron pyrites in the manufacture of sulphuric acid.

Copper Extraction.

Bearing in mind the varied composition of the ores, it will be seen that the processes of extraction will vary much according as to whether one or several kinds of ore are to be treated together. Thus the reduction of oxidized ores only is a simple matter, but when sulphuretted and oxidized ores must be treated together, the operation is more involved. This is the case in this country, where the supply of oxidized ores is insufficient to warrant their separate treatment.

Treatment of Sulphides with or without the Addition of "Oxidized Ores of Copper."

Reaction Process.—Sulphides, with the exception of copper glance, do not usually contain sufficient copper to permit of its direct extraction, and undergo a series of operations, the object of which is to concentrate the copper in a rich regulus. These operations consist of alternate roastings and fusions in a *reducing* atmosphere. In the roasting processes sulphur and arsenic are oxidized and removed as sulphur dioxide (SO_2) and arsenious oxide (As_2O_3) respectively, and the iron and copper are partly oxidized.

$$Cu_2S + 3O = Cu_2O + SO_2$$
$$Fe_2S_3 + 9O = Fe_2O_3 + 3SO_2$$
$$FeS_2 + O_2 = FeS + SO_2.$$

In the fusion which follows, the reaction of the oxide of copper on the remaining sulphide of iron produces sulphide of copper and oxide of iron. At the same time, many impurities are removed. The iron oxide thus produced, together with that formed during roasting, is removed by combining with silica, to form silicate of iron, which constitutes the slag. There is always sufficient silica in the charge and furnace bottom to effect this.

$$2Cu_2O + 2FeS + SiO_2 = 2Cu_2S + 2FeO.SiO_2$$
$$Fe_2O_3 + CO + SiO_2 = 2FeO.SiO_2 + CO_2.$$

The sulphide of copper and unaltered iron sulphide fuse and form a bottom layer in the furnace. This process is repeated until the iron has been practically removed, and the regulus is then treated for copper by roasting and fusion, and afterwards refined. Oxidized ores and the slags produced in the fusions, which contain too much copper to throw away, are introduced in the fusions, and the metal they contain by reaction on the sulphide of iron passes into and enriches the regulus.

(1) **Welsh Process.**—This is conducted in reverberatory furnaces throughout. The ore mixture contains from 9 to 13 per cent. of copper as sulphide, with excess of iron pyrites and silica, and the process involves at least six operations—

(*a*) Calcining the ore;
(*b*) Fusion of the calcined material with oxidized ores and slag;
(*c*) Calcining the regulus obtained in *b*;
(*d*) Fusion of calcined regulus with slags;
(*e*) Roasting and fusion of regulus with separation of blister copper;
(*f*) Refining and toughening.

(*a*) **Calcining the Ore.**—This is conducted in a large reverberatory furnace, the bed of which is shown in Fig. 56. The bed measures about 16 feet by 14 feet and the grate 4 feet by 3 feet. Air is admitted at the fire-bridge through openings *o*. The temperature is low, and the ore is turned over from time to time. About half the sulphur is removed mainly as sulphur dioxide SO_2. Some sulphur trioxide is also evolved. Arsenic passes off as As_2O_3. This roasting on 3-ton charges occupies about 24 hours. The charge is introduced from hoppers on the roof of the furnace. The roasted ore is raked through openings *r* in the bed, into the vault below, which communicates with the flue, to cool.

(*b*) **Fusion for "Coarse" Metal.**—The roasted ore is mixed

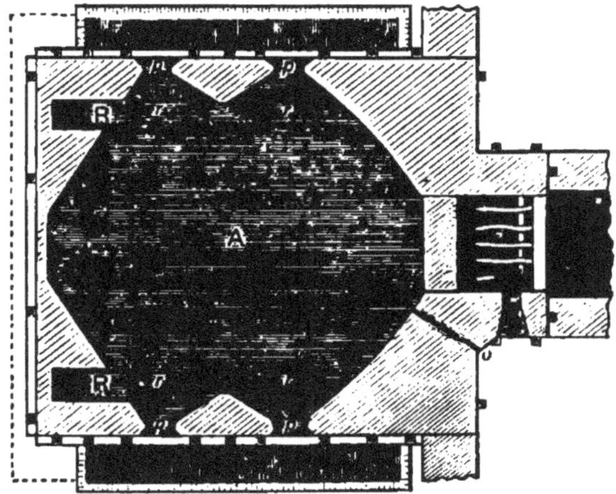

FIG. 56.—Plan of Bed of Calcining Furnace. A, bed; F, fireplace; *p*, doors; *r*, openings into vault.

with oxidized ores and slags, the charge being made up as follows:—

Roasted ore, 60 to 66 per cent.

Oxidized ores, 10 to 14 per cent.

Metal-furnace slag from fourth process, 22 to 25 per cent.

This is introduced in charges of about 25 cwts. into the "ore furnace" (Fig. 57). This is also a reverberatory furnace, but the grate area is much larger in proportion to the bed, a higher temperature being necessary to fuse the charge. The furnace is charged from the hopper A. The bed is of sand, and slopes from all points towards the tapping hole B, under the door C in front of the furnace. The charge gradually fuses, the oxides, sulphides, and sulphates react as described, and the regulus formed separates and collects at the bottom under the slag. This slag—*ore-furnace slag*—often contains unfused masses of quartz and stony matters occurring in the ore, and is raked off the surface of the regulus through a long low opening D, at the flue end of the furnace into sand moulds, E, beneath. Before tapping, three charges are generally fused, and the whole of the regulus is run out together into sand moulds. A more uniform product is thus obtained. Sometimes the regulus is granulated by tapping into a tank containing water.

The metal-furnace slag—sharp slag—which forms part of

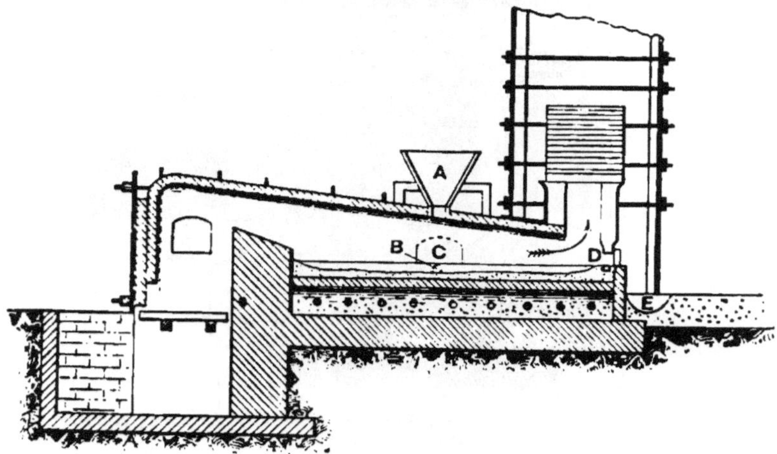

FIG. 57.—Ore Furnace.

the furnace charge, contains about 4 per cent. of copper as silicate, etc. This reacts upon the iron sulphide present, and the copper passes into the regulus, thus affording a convenient method of treating the cupreous slags produced in the several operations.

It is essential that the furnace charge should contain an excess of sulphide of iron to ensure the complete decomposition of the oxide of copper in the roasted material and in the oxides, carbonates, slags, etc., added.

The regulus obtained consists essentially of a mixture of iron and copper sulphides, containing 30 to 35 Cu, 30 Fe, 28 S, with small quantities of arsenic, bismuth, lead, antimony, and sometimes, tin, nickel, and cobalt sulphides. It is known as "coarse metal," and breaks with a coarsely granular fracture of a bronze-purple colour. The slag is known as "*ore-furnace slag*," and consists principally of silicate of iron, and contains less than 1 per cent. of copper.

(*c*) **Calcining Coarse Metal.**—The pigs of coarse metal are crushed, unless it has been granulated, and the regulus is roasted in the calciner for 24 hours at a low red heat, losing about one-half its sulphur, mainly as SO_2.

(*d*) **Fusion for "Fine" Metal.**—The calcined coarse metal is mixed with roaster and refinery slags (slags from the fifth and sixth processes, containing a large proportion of cuprous oxide as silicate), and also sometimes with pure oxide and carbonate ores.

The mixture contains—

Roasted regulus, 65 to 80 per cent.
Slag and oxidized ores, 20 to 35 per cent.

The "metal furnace" in which the fusion takes place is similar to that employed in this fusion for "coarse metal." The charge consists of about 30 cwts., and its fusion occupies from 6 to 8 hours. The same reaction between the sulphides and oxides in the charge occur as before.

The second calcination and fusion have for their object the production of a rich regulus as free as possible from iron. The extent to which this has been accomplished depends on the efficiency of the roasting and the amount of oxidized cupreous materials added. If the oxide of copper is insufficient to decompose the sulphide of iron present, a regulus, which breaks with a smooth, shining fracture and a bluish colour, containing from 55 to 66 per cent. of copper, known as "*blue metal*," results. It is a mixture of cuprous and iron sulphides.

When the oxide of copper is in the required proportion, a regulus which breaks with a semi-metallic, greyish-white, slightly granular fracture, known as "*white metal*," is produced. It contains from 70 to 78 per cent. of copper, and is practically "cuprous sulphide," Cu_2S. "*Pimple metal*" contains a larger percentage of copper, and is produced when the oxide of copper is in excess.

Sometimes the oxide is in excess of requirements, in which case it reacts on the sulphide during fusion, and metallic copper is produced with an evolution of SO_2. The regulus appears to dissolve a certain amount of the reduced copper, which separates out, as the metal cools, in fine, velvety filaments, lining cavities in the metal, and is known as "*moss copper.*" Separated copper is also found in blue metal, but is absent in pure white metal.

"*Metal-furnace slag*" has a bluish, lustrous, semi-crystalline fracture. It is essentially silicate of iron, but contains about 4 per cent. of copper, which is recovered in the second operation.

(*e*) **The Roaster Stage.**—The pigs of fine metal are placed on the bed of the roaster furnace, which in most respects resembles the metal furnace, but is, however, provided with openings at the fire-bridge, for the admission of air into the furnace chamber, and a basin-shaped depression in front of the door. The temperature is so managed that the melting down occupies from 6 to 8 hours. Extensive oxidation takes place, sulphur passing off as sulphur dioxide; thus—

$$Cu_2S + 3O = Cu_2O + SO_2.$$

When melted, the slag which has formed is skimmed off, and the clear surface presents a boiling appearance, and emits a frizzling sound, due to the escape of SO_2 formed by the oxide reacting on the sulphide.

$$Cu_2S + 2Cu_2O = 3Cu_2 + SO_2.$$

Metallic copper separates and sinks to the bottom of the bath. When the process is judged complete, the slag is again skimmed off and the copper tapped into sand moulds. The length of time occupied varies from 12 to 24 hours, being longest when the fine metal is least pure.

The "*blister copper*" is dry and unmalleable, has a dull, red fracture, and contains cavities. The surface presents a blistered appearance, caused by sulphur dioxide liberated during solidification; hence the name. It contains about 98 per cent. of copper and less than 1 per cent. of iron.

Roaster slag has a purplish red colour, and contains from 17 to 40 per cent. of copper, as silicate and metal, according to circumstances.

(*f*) **Refining and Toughening.**—The refining furnace has a sand bottom, which inclines from all parts to a basin-shaped cavity near the end door. There is also no charging hopper or tap-hole. Some 6 to 15 tons of pigs of blister copper are piled up on the bed, and gradually melted. This occupies from 4 to 6 hours. The slag is skimmed off, and the surface exposed to the oxidizing atmosphere for from 10 to 15 hours longer. Copper being less oxidizable than the impurities present, viz. arsenic, sulphur, iron, tin, nickel, cobalt, manganese, bismuth, antimony, and lead, these are removed as oxides; but much copper, owing to its great excess, is also oxidized, forming cuprous oxide. This, with the other metallic oxides formed, combines with silica from the sand bottom, etc., and constitutes the slag. Some cuprous oxide is, however, dissolved by the metal, and renders it "dry" or "underpoled." The attainment of this state is determined by the withdrawal of samples from the bath. To remedy this, the slag is again skimmed off, the surface of the copper covered with coal or anthracite, and a pole of green birch or oak wood is plunged into the metal and held down. The hot metal causes a copious evolution of steam and reducing gases, which thoroughly agitate the metal, bringing every portion of it into contact with the carbonaceous matter covering it, whereby the cuprous oxide in the metal is reduced. Samples are taken from the bath from time to time, examined, and tested for toughness and malleability. When the metal has lost its dark-red, granular fracture, and breaks with a flesh-coloured silky lustre, bending double when placed in a vice, the "tough pitch" has been attained: the pole is withdrawn, the covering pushed aside, and the metal ladled out by hand, in clay-covered ladles, and cast into flat ingots weighing about 20 lbs. The ingot moulds are of cast iron or copper, and so arranged that, as soon as solid, the ingots can be thrown into water. During ladling the metal is apt to become oxidized and dry again. When this is observed, the pole is reintroduced for a short time, and the metal brought back to toughness. The time occupied is about 30 hours, from introduction of pigs to end of ladling out.

Refinery slag is of a coppery-red colour. It consists mainly of silicate of copper, not unfrequently with shots of metal freely dispersed through it.

Modifications of the Welsh Process.—In some cases, owing to poorness of the ores, or the lack of oxides and slags, or sometimes difference in practice, the number of operations is increased, the roaster stage for the production of blister copper being preceded by more calcinations and fusions, to produce a satisfactory regulus.

A small quantity of lead is often added to copper intended for rolling, just before ladling; after which the scum of oxides which forms is skimmed off. The amount added varies from 0·1 to 0·5 per cent. Its object is twofold. By its oxidation it promotes the oxidation and removal of foreign metals present, notably antimony, and also retards the metal from "going back" by oxidation, and again becoming dry and underpoled. The ingots are sounder and flatter. Copper containing not more than 0·1 per cent. of lead rolls well, and has a less tendency to collar the rolls. It is a little more difficult to detach the scale. No addition of lead is made in making best selected copper, or copper intended for making best brass, gun metal, or German silver.

"**Best selected**" **copper** was formerly made from the purer portions of the fine metal resulting from the smelting of purer ores. It is found that the impurities become concentrated by gravity in the lowest part of the bath, and consequently those pigs which flow from the furnace first are most impure. The selection of the later ones for making the best copper led to the introduction of the term. Best selected copper should contain only traces of arsenic, antimony, and bismuth.

Another process of selection is now followed, known as the "bottoms" process. The "fine metal" obtained in the fourth stage, prior to tapping, is cleared of slag and roasted. The oxide of copper formed reacts on the sulphide, and copper is reduced. This attacks the foreign sulphides, reduces them, and alloys with the liberated metal, thus concentrating them in the metallic state, and carrying them to the bottom of the bath. This leaves the "fine" metal purified. The "bottoms" copper contains nearly all the gold and silver, and much of the tin, lead, and antimony in the charge.

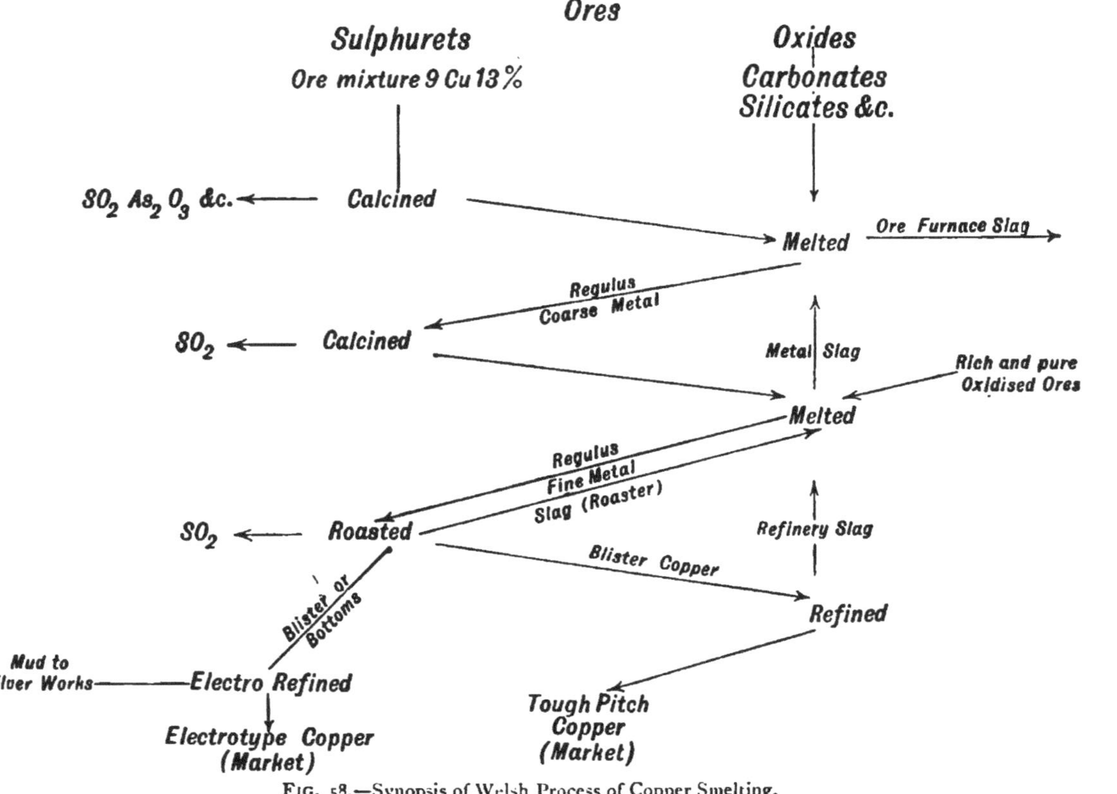

FIG. 58.—Synopsis of Welsh Process of Copper Smelting.

Instead of employing reverberatory furnaces exclusively, blast furnaces of the "water-jacketed" type are commonly used abroad for the fusion of ore and regulus, and the roasting is done in heaps or in stalls (see p. 25).

Reduction Processes.—Oxides, carbonates, and other oxidized copper ores, if in sufficient quantity, may be smelted in blast furnaces—preferably water-jacketed—with coke and a suitable flux, *e.g.* oxide of iron, to slag off silica. By adding a little iron pyrites to the charge, a small amount of copper regulus is formed, as well as metallic copper, and the slags are cleaned.

Sulphide of copper is not completely reduced by iron or carbon. It is necessary, therefore, to convert "mattes" into oxide, before reduction, by calcining. They may then be treated as above. This is also done with the "fine regulus" obtained at Mansfeldt in Germany, where the regulus is desilverized by the Ziervogel process. The finely divided residue of oxide of copper and iron is balled with a little clay and smelted for "black" copper, which is afterwards refined by an air blast on a small hemispherical hearth, under charcoal.

Mannhes Process for Bessemerizing Copper Mattes.—Successful attempts have been made to treat copper matte on the Bessemer principle, by blowing air through it in the molten state. The sulphur is burnt out and iron oxidized, and sufficient heat developed to keep the mass molten. It is found advisable to treat the matte in two stages, producing fine metal and coarse copper in distinct operations. The converter is more or less barrel-shaped, and the tuyeres enter at some distance from the bottom, in a more or less horizontal direction. By this arrangement the enriched matte or copper falls below the tuyere level, and escapes further oxidation. It is run into the converter at as high a temperature as possible. There is a considerable saving in fuel in this method of treatment.

In a few cases the matte is Bessemerized in the reverberatory furnace, by blowing air through the bath of melted sulphides, movable tuyeres being employed for that purpose. By this means the sulphur is burnt out and iron oxidized and removed, all the reactions of calcination and fusion taking place in the fluid bath.

Direct Process.—At Briton Ferry, working with ores of special purity, the "roaster stage" is conducted in a totally different fashion from that generally prevalent. A portion of the fine metal is roasted "sweet" in a calciner of the revolving type (Fig. 17), and mixed with a sufficient amount of the unroasted regulus—determined by experiment—to reduce it. The mixture is then heated to fusion in a reverberatory furnace. A much greater yield of copper is obtained in this way, and its quality is said to be in no way inferior. The copper is at once refined in the same furnace.

"Basic" linings have been introduced in the roaster and refining furnaces. The loss of copper in the slag is much less, owing to the absence of silica, which by combining with the oxide formed in roasting retards its reaction on the sulphide. The yield of blister copper is said to be 25 per cent. more than with sand bottoms. Arsenic is removed to a greater extent, but bismuth and antimony in no greater degree than on a siliceous bottom. In refining arsenical metal, soda ash is added as well as lime.

Chili Bar, American Matte, etc.—In districts where fuel is dear the ores are subjected to a preliminary roasting, and then fused in water-jacketed furnaces for a matte running about 45 per cent. of copper. This is then roasted as completely as possible, and remelted with the production of blister copper and cuprous sulphide (white metal). The copper contains about 1 per cent. sulphur, in addition to the usual impurities. This is marketed as Chili bar.

Wet Processes of Copper Extraction.

In wet methods of extracting copper the metal must first be converted into a soluble form, as sulphate or chloride, and the copper deposited from solution by scrap iron.

Sulphate Roasting.—The conversion of the copper in pyritical ores into sulphate may be effected by careful calcining in a reverberatory furnace at a low heat. The sulphide of copper is converted into sulphate, partly by direct oxidation, thus—

$$\underset{\text{Cuprous sulphide}}{Cu_2S} + \underset{\text{Oxygen}}{5O} = \underset{\text{Copper sulphate}}{CuSO_4} + \underset{\text{Cupric oxide}}{CuO}$$

$$\text{or } \underset{\text{Cuprous sulphide}}{2\,Cu_2S} + \underset{\text{Oxygen}}{7O} = \underset{\text{Copper sulphate}}{CuSO_4} + \underset{\text{Cuprous oxide}}{Cu_2O} + \underset{\text{Sulphur dioxide}}{SO_2}$$

and partly by the SO_3 liberated from ferrous sulphate formed by the calcination of the iron sulphide present, or produced by the combination of oxygen with the SO_2 generated, brought about by the "contact action" of the ferric oxide and silica, and brickwork of the furnace—

$$\underset{\text{Iron pyrites}}{FeS_2} + \underset{\text{Oxygen}}{6O} = \underset{\text{Ferrous sulphate}}{FeSO_4} + \underset{\text{Sulphur dioxide}}{SO_2}$$

$$\underset{\text{Ferrous sulphate}}{2FeSO_4} + \underset{\text{Oxygen}}{O} = \underset{\text{Ferric oxide}}{Fe_2O_3} + \underset{\text{Sulphur trioxide}}{2SO_3}$$

$$\underset{\text{Sulphur dioxide}}{SO_2} + \underset{\text{Oxygen}}{O} = \underset{\text{Sulphur trioxide}}{SO_3}$$

$$\underset{\text{Copper oxide}}{CuO} + \underset{\text{Sulphur trioxide}}{SO_3} = \underset{\text{Copper sulphate}}{CuSO_4}$$

Copper sulphate requires a higher temperature to decompose it than ferrous sulphate, but is decomposed more readily than silver sulphate.

There is great difficulty in getting the whole of the copper sulphated. A greater amount is rendered soluble in the presence of much sulphide of iron.

Bankart's and Escalle's processes, both now abandoned, were based on this principle. In the former the copper was precipitated by iron, and in the latter as sulphide by calcium sulphide. This sulphide was subsequently reduced in a special form of furnace, and refined.

Much copper is extracted by calcining "low grade" pyritical ores in open heaps, washing out the sulphate formed in tanks, and precipitating the copper by iron.

Certain sulphide ores oxidize spontaneously on exposure to moist air. From this cause the water from copper mines, and drainage from cupreous waste heaps, often contain copper sulphate in solution. Extensive works were executed for precipitating mine waters in the Carnon Valley in Cornwall, at Pary's Mountain in Anglesea, and elsewhere. Both these methods are followed at the Rio Tinto mines.

Chloridizing Processes.—The conversion of copper into chloride is effected by roasting sulphide ores with salt (sodium chloride), or by treating with some chlorinating agent, such as ferric chloride or manganese dioxide and salt, which, in the presence of sulphuric acid or sulphates, generate chlorine and hydrochloric acid.

Chlorinating Roasting.—In roasting with salt, the sulphates produced react on the salt and form sulphate of soda.

$$CuSO_4 + 2NaCl = CuCl_2 + Na_2SO_4.$$

Chlorine and hydrochloric acid are also generated in the furnace (see p. 221). The chlorine in the salt is ultimately transferred to the copper, which is converted into cupric and cuprous chlorides, the former soluble in water, and the latter in hydrochloric acid and chlorides.

Longmaid and Henderson's Processes.—This process is adopted for the treatment of the cinders from the burning of iron pyrites used in the manufacture of sulphuric acid. Portuguese, Spanish, and Norwegian pyrites, so largely imported for this purpose, contain from 1 to 2·5 per cent. of copper, which, after burning off the sulphur, reaches from 2 to 5 per cent. The "purple ore," as it is called, is ground down and mixed with a little small green ore (unburnt pyrites) and 10 to 18 per cent. rock salt in a mechanical mixer. This is roasted at a very low temperature, between 400° and 500° C. (copper chloride being volatile at a high temperature), for about 8 hours in a reverberatory or close muffle furnace (see p. 28). The roasted ore after withdrawal is lixiviated in wooden tanks, first with water and then with hydrochloric acid, obtained by passing the furnace gases up through condensing towers where the hydrochloric acid generated in roasting is dissolved out. From these it is run into settling tanks placed at a lower level, and then into precipitating tanks, where the copper is thrown down by iron. Generally, however, the ores treated contain gold and silver, and these are also extracted (see Claudet's Process for Silver). The "copper precipitate" is collected, fused, and refined.

In chlorination in heaps, the ore, part of which is calcined, is stacked in huge heaps mixed with salt, manganese dioxide, and residues from previous heaps. Open channels are left for the admission of air and moisture. Decomposition sets in with the production of ferric and manganese chlorides, which chlorinate the copper. The heaps are periodically drenched with water; or the heaps may be drenched with water containing ferrous chloride, etc., produced in the precipitation of the copper, to hasten the process. A series of chemical reactions—somewhat involved—result in the conversion of the copper into chloride. The copper solution is precipitated by scrap iron.

Electro refining of Copper has made great strides in consequence of the demand for pure copper for electrical work. The copper to be refined is cast in thick plates, which are enveloped in canvas bags. They are connected with the positive pole of a dynamo, and immersed in a bath consisting of a 15 per cent. solution of copper sulphate and 5 per cent. sulphuric acid, alternating with thin plates of pure copper connected with the negative pole. When the current passes copper is deposited on the thin plates, and the acid liberated at the + pole attacks and dissolves the copper, which in turn is deposited. The impurities it contains either remain dissolved in the bath, or are left in an insoluble form as a mud, which is retained in the bags. This method of refining is largely followed for argentiferous "bottoms" copper, the silver and gold remaining in the insoluble residues.

Varieties of Commercial Copper.

Tough Cake, or Tough Pitch Copper is ordinary copper at its point of greatest malleability and toughness.

Bean Shot Copper and Feather Shot for brass making are made by pouring molten copper into *hot* or *cold* water. For this purpose the copper is "overpoled."

Rosette Copper is obtained in thin films of a fine red colour by throwing water on the surface of the metal when molten, and lifting off the solidified crusts.

Chili Bar is imported in bars weighing about 2 cwt. It is somewhat less pure than blister copper, and requires refining.

Copper Precipitate is the finely divided copper obtained by precipitating copper from solutions by iron. Its purity is very variable. The foreign matter is principally oxide of iron.

Electrotype Copper.—Electro deposited, or electro-refined copper is produced by electro deposition, using "bottoms" or other impure copper as the anode or dissolving pole.

CHAPTER XIII.

LEAD.

Physical Properties.—The metal possesses a bluish grey colour, and considerable lustre on fresh surfaces, which, however, are soon dimmed on exposure. It is soft enough to be impressed by the nail and to mark paper. Impurities, such as antimony, render it harder. It is malleable, ductile, and tough, but is very deficient in tenacity. Cast lead has only a tenacity of from 0·4 to 0·8 tons per square inch. After wire-drawing this is increased to 1 to 1·75 tons. Its melting-point is about 330° C., and it volatilizes at very high temperatures. It contracts on solidifying, and is consequently unsuitable for castings. The specific gravity is 11·36, and is not increased by hammering. When alloyed with other base metals, the specific gravity is diminished. It welds readily if the surfaces are fresh and clean, and even lead powder may be moulded by pressure into solid lumps. Alloys with tin may also be thus produced, and a compound sheet of the two metals may be formed by placing them in contact and passing through the rolls. The flowing power of the metal is great, and lead pipes and rods are squirted from a press. Lead crystallizes on cooling from fusion. When heated near its melting-point, it breaks with a columnar fracture.

Chemical Properties.—Lead oxidizes on exposure to moist air, forming suboxide of lead (Pb_2O). In a very finely divided state, as obtained by heating the tartrate, it takes fire and burns. When heated in air it readily combines with oxygen, and forms lead monoxide, litharge (PbO). This oxide is of a yellow colour; it fuses at a full red heat, and yields, on cooling, a yellow crystalline mass. At a somewhat higher temperature, it combines with silica, forming a readily fusible silicate of lead.

On this account it rapidly corrodes crucibles, retorts, and the sides of furnaces made with siliceous materials. Hence the necessity of employing bone ash or marl brasque in cupellation (p. 225), and the advantages of using water-jacketed furnaces in smelting operations. Mixtures of cuprous and lead oxides are even more corrosive than litharge alone. Litharge is largely used in glass making, and is made by oxidizing lead on a cupel. See page 225.

It exerts an oxidizing influence on iron, copper, zinc, and other metals, being reduced to lead. When heated with oxides of other metals, such as copper and iron oxides, litharge fuses and dissolves up the refractory oxide, forming a fusible mass. The amount of litharge required varies. Thus 1 part of cuprous oxide requires 1·5 parts of litharge, while 1 of tin oxide requires at least 12.

If produced at a temperature below its fusion point, litharge has a brownish-yellow colour, and is known as *massicot.* If this is carefully heated in air, it takes up more oxygen, and is converted into red lead or *minium* (Pb_3O_4).

Manufacture of Red Lead.—The metal is first "drossed," or oxidized in a low, reverberatory furnace, or "oven," with two narrow fireplaces, one on either side of the bed. The products of combustion escape through the working door in front, and are carried away by a hood surmounted by a chimney. The bed of the oven slopes slightly to the middle, and from back to front. In working, a dam of rough oxide, mixed with lead from the grinding of previous charges, is made across the front of the oven, and some 20 to 30 cwts. of lead charged in and melted at a low red heat. The door is left partly open, and the oxide as it forms is pushed back, and the lead splashed about by a long iron paddle. The metal is continually thrown over the oxide at the back. Oxidation goes on freely, and the unoxidized lead drains to the front. Drossing is assisted by the addition of a little antimony to the lead. When the oxidation is completed, the charge is raked out into iron barrows and allowed to cool. It is then ground by millstones in a stream of water, which carries the fine material away in suspension. The unoxidized metallic lead and heavy particles of oxide are left behind in the troughs which lead to the settling tanks, where the finely divided massicot is deposited. It is collected and dried. It then constitutes "ground litharge." This is transferred to the "colouring oven," very similar to the drossing oven, except that the bed is flat. It is spread in low ridges over the bottom of the oven and "coloured" at a lower temperature than that used in the drossing operation, being turned over from time to time. The red lead while hot has a deep brownish purple colour, and is examined from time to time by the withdrawal of samples, and allowing to cool. When the oxidation is complete, the cold sample has a bright red colour. It is again ground, levigated, and after drying and sieving, packed in barrels. Its composition is Pb_3O_4. On heating it gives off oxygen and forms litharge, PbO. Treated with nitric acid, PbO_2, lead peroxide is left as a purplish powder.

Action of Soft Water on Lead.—Waters containing oxygen in solution readily attack lead, but the action is retarded by the presence of carbonates and sulphates in the water. Lead pipes for conveying soft waters are coated inside with tin to prevent the water from being contaminated with lead.

Lead and Sulphur.—Lead combines readily with sulphur when heated, forming a brittle, grey, crystalline mass of lead sulphide (PbS). It has a high metallic lustre, and *melts at a higher temperature than the metal.* At a full red heat it is decomposed by iron, sulphide of iron and metallic lead resulting; thus—

$$2PbS + Fe_2 = 2FeS + Pb_2$$

When calcined it is partly converted into oxide and partly into sulphate, SO_2 passing off.

The sulphate, which is also produced by the addition of sulphuric acid to a soluble salt of lead, is a white substance not readily decomposed by heat. It is insoluble in water. Heated with carbon, it is reduced to sulphide.

When sulphide of lead is heated with oxide or sulphate,

the sulphur and oxygen combine and pass off as SO_2, metallic lead separating thus—

$$2PbO + PbS = 3Pb + SO_2$$
$$PbS + PbSO_4 = Pb_2 + 2SO_2.$$

Lead Ores.—The principal ores of lead are the sulphide, carbonate, and chlorophosphate.

Galena, *blue lead ore, lead sulphide* (PbS), is the most important and abundant. It is found both crystalline and massive. It has a grey metallic lustre, and is heavy, having a specific gravity of about 7·5. It is brittle, and contains 86·6 per cent. of lead. Galena occurs widely distributed in the older rocks. It is usually associated with quartz, fluor, calcite, barytes, and spathic iron ore in the veins, and frequently with copper pyrites and zinc ores. It often contains silver, sometimes in considerable quantity. Such ores are described as argentiferous. Iron, antimony, copper, and zinc are commonly present, and gold and bismuth also frequently occur in it. The localities are very numerous.

Cerusite, *lead carbonate,* or *white lead ore* ($PbCO_3$), also occurs. Its colour is white or yellowish, and its lustre adamantine to earthy. It has a specific gravity of 6·5, and contains 75 per cent. of lead. It is frequently argentiferous, like galena. The deposits at Leadville, in Colorado, and at Broken Hill, in Australia, are of this character.

Anglesite, *lead sulphate* ($PbSO_4$), also occurs, associated with galena and other lead ores.

Pyromorphite, *green lead ore, linnets, chlorophosphate of lead* ($3Pb_3(PO_4)_2, PbCl_2$) occurs in hexagonal crystals and as green and brown masses. Its specific gravity varies from 5·5 to 7·2. Ores in which the phosphorus is replaced by arsenic are known as *mimetesite.* In addition to the above, many compounds of lead occur naturally, among which may be mentioned Boulangerite ($3PbS, Sb_2S_3$) and Jamesonite, another antimonial sulphide of lead.

Lead Smelting.

So many lead ores contain silver that the metallurgical treatment of the two metals is hardly separable. In this chapter we purpose dealing with the extraction and refining of lead, and such processes for the concentration of the silver as are conducted upon the lead, and to leave the actual recovery of the silver to be dealt with when treating of that metal. Extraction processes may be grouped, in much the same manner as in copper smelting, into reaction and reduction processes.

The *reaction* process for gelena is based on similar chemical changes to those which occur in copper smelting, viz. the mutual reaction of the unaltered sulphide upon oxide and sulphate formed by roasting the sulphide. In the case of lead, however, the operation is simplified. The ore as received from the miner contains a sufficient percentage of metal to permit of its direct treatment.

The *reduction* processes may be divided into carbon reduction, where that element forms the reducing agent, and iron reduction processes, where iron, or iron bearing materials, such as oxides of iron, or iron slags, form part of the charge, and liberate the lead from combination.

Reaction Processes.—To this category belong the Flintshire, Derbyshire, Spanish, French, and Bleiberg methods of smelting galena.

The form of furnace employed and the details of the process vary greatly, having regard to the purity of the galena or its admixture with carbonate, sulphate, etc.

The **Flintshire** furnace is shown in Fig. 59. It is a reverberatory furnace, having three doors opening on either side of the hearth. The side on which the firing door is situated is known as the "labourers' side," and that opposite as the "working side." The bed—which consists of slag from previous operations, spread over the hearth while in a pasty state—is level with the doors on the labourers' side, but, on the working side, slopes so as to form a well some 18 inches deep, immediately in front of the middle door. A tap-hole, B, communicating with the bottom of this, is provided for tapping out the lead. A second tap-hole above this, for the removal of fused slag, is provided in some furnaces. Outside the furnace is an iron pot, into which the metal is tapped. At the top of the furnace is a hopper, from which the ore is introduced into the furnace.

The process is conducted as follows. The charge of about a ton is let down from the hopper into the furnace, still red hot from a previous charge, and spread from the labourers' side over the furnace bed, clear of the well. It is then calcined at dull redness for from one and a half to two hours, being stirred and turned over from time to time to expose it to the air, to admit which the doors are left partly open. The temperature is not sufficient to melt the galena, which, it may

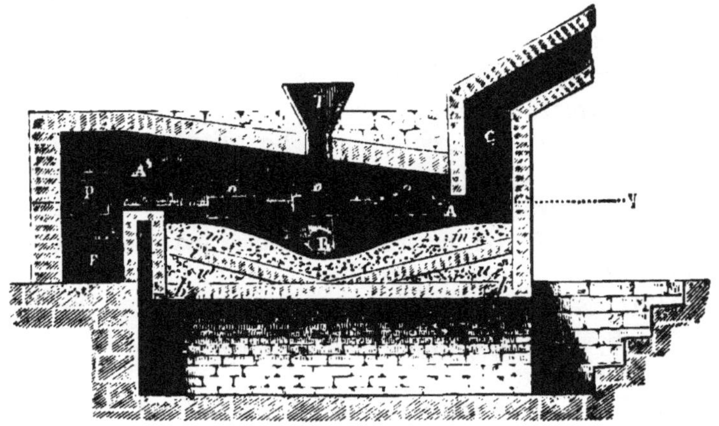

Fig. 59.—Lead Smelting Furnace.

be noted, has a higher melting-point than lead itself. During this stage oxidation occurs freely, oxide and sulphate of lead being formed.

The doors are now closed, the fire made up, and the temperature raised to full redness, when the reaction between the oxide, sulphide, and sulphate produce a copious separation of lead, which collects in the well of the furnace. The temperature at this point is also somewhat below the melting-point of galena.

The unreduced mass becomes soft and pasty. It is pushed out of the basin and spread over the hearth. To prevent its fusion, the doors are opened to cool it somewhat, and it is stiffened—"set up"—by the addition of a little lime.

The charge is then melted—"flowed"—down by increasing

the temperature, for which purpose the doors are closed, the damper opened, and a fresh fire made. The lime decomposes any silicate of lead formed, producing silicate of lime, and liberating oxide of lead. This, reacting on the unreduced sulphide present, a further separation of lead occurs. Lime is again thrown in and mixed with the slags to render them pasty, and they are again spread over the hearth and roasted for from half to one hour. At the end of this period the temperature is raised to its highest point, and the whole is melted a second time. The oxide produced during the roasting and that liberated from silicate by the lime added is in this stage often more than sufficient to decompose the remaining sulphide, and a little coal slack is often added to assist in its reduction. This also reduces any sulphate to sulphide, which reacts on the oxide, producing lead. The metal is then tapped into the lead-pot in front.

The slags are dried by further additions of lime, and are withdrawn in a pasty state from the furnace. They are known as grey slags, and usually amount to about 20 per cent. of the charge. They contain about 40 per cent. of lead as silicate, which is recovered in slag hearths.

Setting up by lime has a twofold object. Its principal use is to stiffen the slags and render them infusible during the roasting periods, so that the galena contained in the charge shall not become bound up, and protected from the action of the air. In the fusions it probably liberates oxide of lead from the silicate.

The metal in the lead-pot is covered with slags, matte, and dross, which retains much metallic lead in globules. Coal slack is thrown on top and stirred into the hot metal. The gas produced burns on top, heats the slag, and releases the shots of metal. The skimmings are either thrown back into the furnace at once, to further separate lead, or are added to the succeeding charge near the end of the preliminary calcining.

The stages of the process are known as "fires."

When barytes occurs as gangue in the ore, it is necessary to add fluor spar as a flux, or to mix it with ore containing fluor. The amount of blende in the ore also influences the fusibility of the slags.

The processes in use at Cuëron, Blieberg, and elsewhere

are of a similar character. This process is only applicable to pure ores. Foreign sulphides such as antimonite and even copper pyrites combine with galena, and form readily fusible double sulphides, which melt or clot, and arrest the roasting.

In smelting lead ores containing antimony in reverberatory furnaces by the reaction process, the lead obtained in the earlier stages is freer from that element than that subsequently produced.

Reduction Processes.—These are conducted in both reverberatory and blast furnaces. They are employed for the treatment of impure ores and slags, and for reducing the oxides, dross, or *abstriches* produced in the purification of lead.

In the treatment of raw ores, iron is the reducing agent generally employed. With poor ores containing much iron sulphide, a preliminary roasting and fusion are resorted to, for the purpose of fluxing off the iron and concentrating the lead.

The Cornish Process.—This process is followed to some extent for the treatment of impure ores containing copper and antimony, reguli containing lead, and also for the treatment of slags.

The ore or regulus is first roasted in a separate calciner, much as in copper smelting, for from 15 to 18 hours.

It is then smelted in a furnace resembling a Flintshire furnace. The charge of 2 tons is melted down in from 2 to 3 hours.

With pure ores, or with substances rich in silver, the lead which separates by "reaction" is removed and dealt with separately. In the former case it is purer, and in the latter richer in silver than that subsequently produced.

Lime and anthracite culm are then added and well mixed. The materials, thus rendered stiff, are spread over the hearth, and some 2 cwts. of scrap iron added. The doors are closed and luted, and the charge remelted at a high temperature. The products separate in layers, which follow each other when the furnace is tapped. *Lead*, which is received in the lead-pot in front; *regulus*, or slurry, which is a mixture of iron sulphide, with the copper and some lead sulphide which flows over the top of the lead-pot into the pot below; and *slag*, which is generally so free from lead and copper as to be thrown away.

The process occupies about 8 hours.

In this process the lead first obtained is the result of reaction between oxide, sulphide, and sulphate; in the second stage the sulphide and silicate present are reduced by the iron. The anthracite reduces the oxidized matters present.

$$2PbO,SiO_2 + 2Fe = 2FeO,SiO_2 + Pb_2.$$

Lead Smelting in Blast Furnaces.—Blast furnaces are now largely employed in lead smelting, water-jacketed furnaces (Fig. 60) being principally employed. The ore, unless an oxidized one (carbonate, phosphate, etc.), is first roasted in a reverberatory furnace, being finally heated till it clots together. It is then mixed with iron-bearing materials, such as pyrites cinders from the manufacture of sulphuric acid, iron ores, or puddling and mill furnace slags, which yield metallic iron during the smelting, and suitable fluxing agents, *e.g.* lime. The ore mixture is then smelted with coke as fuel. The oxide of lead is reduced partly by the CO generated, and also by iron reduced by CO in the furnace. $2PbO + Fe_2 + SiO_2 = 2FeO,SiO_2 + Pb_2$.

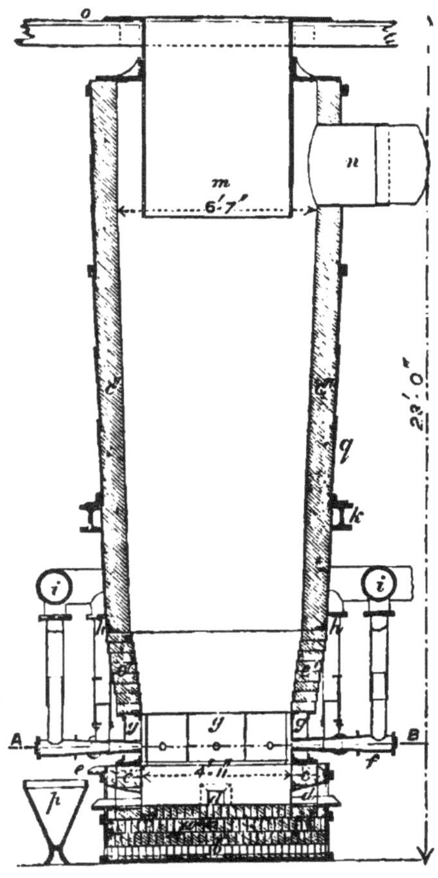

FIG. 60.—*a*, hearth bottom; *b*, channels in brick-work; *dd*, tap holes; *e*, slag lip; *f*, blast pipes; *g*, water-jacket; *i*, blast main; *k*, supporting ring; *m*, charging pipe; *n*, waste gas pipe; *o*, charging floor; *p*, slag pot.

The oxide of iron combines with silica in the charge and passes into the slag.

In these furnaces that portion of the furnace in the vicinity of the tuyeres, which is most seriously corroded by the metallic oxides and slags, when consisting of siliceous materials, is formed of hollow iron casings only, through which water circulates to keep them cool.

Any sulphide of lead is reduced by the iron, iron sulphide being formed.

Three products are obtained :—*Work lead* (containing the greater part of the silver and gold, as well as antimony, tin, bismuth, copper, and traces of cobalt, nickel, and arsenic).

Matte, consisting of sulphide of iron and nearly the whole of the copper in the charge.

It sometimes contains 10 to 12 per cent. of lead and some silver, gold, etc. It is roasted and resmelted in a separate furnace when it yields lead (often rich in silver), a second matte richer in copper, and slag. The second matte is again roasted and smelted, yielding a regulus containing over 20 per cent. of copper and slag. This matte is treated for copper. Sufficient sulphur is left in the mattes after calcining to serve as a vehicle for concentrating the copper. The oxide of iron formed during roasting is fluxed off by the addition of silica in the fusion (see Copper). Slags from the smelting of *first* matte generally contain lead and are resmelted. Lead obtained from matte is very impure.

Slag.—This is essentially silicate of iron, but often contains also notable quantities of lime, alumina, and oxide of zinc ($2FeO,SiO_2 + 2CaOSiO_2$). If lead is present beyond 1 per cent., it is resmelted with the calcined regulus.

It must be remembered that the precious metals have a tendency to associate themselves with the metallic products of an operation. In the above case there are two such products, viz. the lead—which carries the greater part—and the matte which also contains a portion of the precious metals. In the subsequent treatment much of the silver or gold in the matte passes into the lead obtained. What remains ultimately passes into the copper extracted from the concentrated matte, from which it is ultimately recovered.

The Slag Hearth is a small blast furnace used for the treatment of the rich slags obtained in smelting in reverberatory furnaces.

The lead is present in the form of silicate, sulphide, and sulphate, and as much as 40 per cent. is often present. This silicate requires a very high temperature for its reduction by coke. It is more easily reduced by iron.

The slags contain, it must be remembered, a considerable quantity of lime from the "setting up." By mixing them with coal ashes, iron slag, etc. (containing oxide of iron, silica, and alumina), clayey matters (old clay furnace beds or broken brickbats), the alumina and other oxides thus introduced combine with the silica and lime at the high temperature employed, liberating the oxide of lead in the slag, which is reduced by the coke employed as fuel. The lead produced is very impure, and is known as *slag lead.* The slag known as *black slag* is free, or nearly free from lead, and consists of silicates of lime, alumina, iron (hence the colour), and other oxides.

The furnace is shown in Fig. 61. It is rectangular in form, about 26 inches by 22, internal measurement. The hearth itself is 3 feet deep, but it is surmounted with a brickwork hood, or cover, which communicates with flues leading to the stack for the condensation of fume.

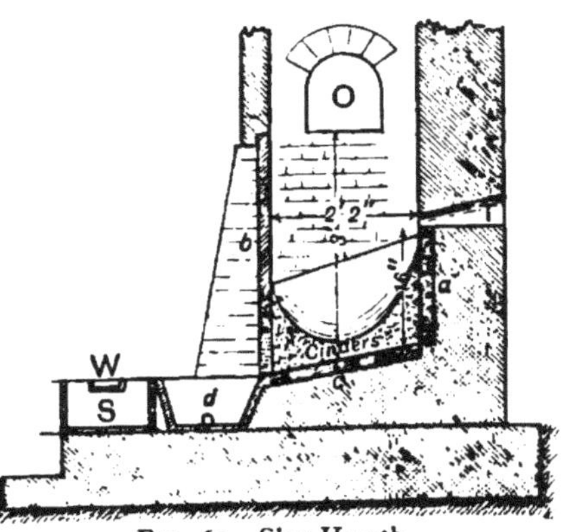

Fig. 61.—Slag Hearth.

The back and side walls of the furnace are built of fire-brick, but below the tuyere at the back is a cast-iron plate (*a*). The front also consists of an iron plate (*b*), the lower edge of which is supported some 7 inches above the cast-iron bed-plate (*c*), leaving an opening across the front of the furnace, stopped with clay while working. The bed-plate (*c*) slopes slightly towards the front to permit the separated lead and slags to flow into the cast-iron receptacle (*d*) in front. This is divided into two unequal parts by a partition which passes nearly to the bottom. The larger compartment is the width of the bed-plate, and is filled with cinders. The lead and slags flow into it from the furnace. The metal filters to the bottom and passes into the smaller division, from which it is ladled while the slags flow over the top of the ashes into the pit (*s*) beyond.

Water flowing through the pit granulates and breaks up the slag, any entangled lead being readily recovered. The tuyere, which is horizontal, enters at the back, and the charging opening is at the side. The lower part of the hearth nearly to the level of the tuyere, and the first compartment of the lead-pot, are filled with cinders, which serve as a strainer for the lead, which flows out from the bottom through openings in the clay stopping. They also protect it from oxidation.

The fire being lighted, coke is introduced, the fire blown up, and the furnace thoroughly heated. Alternate layers of slag and coke are then introduced, and the supply continued as the charge melts down. The slag is removed from time to time by making an opening in the clay breast through the cinder bottom. After working some seven hours, the supply of material is stopped and the fire allowed to burn out. The furnace is cleared out, cooled, and prepared for the next shift.

In the ordinary slag hearth the working cannot be made continuous, as the furnace would get too hot. This would cause serious loss by volatilization, and the furnace walls would be much corroded.

In many works, circular cupolas, with three or more tuyeres and a fore hearth (syphon tap), are employed for the reduction of slags, *e.g.* the Spanish slag hearth and Economic furnace. Water-jacketed, Pilz, and Rachette furnaces[1] are also employed.

Combined Reaction and Reduction Processes.—*The ore hearth*, still in use in Scotland and the north of England, justly holds its own for the production of very pure lead.

The hearth or furnace (Fig. 62) is built of cast-iron plates and blocks, surmounted by a brickwork hood, which communicates with flues.

The bottom consists of a rectangular cast-iron trough, N,—the sump—3 inches thick, measuring about 22 inches square, and some 4½ to 6½ inches deep. This is bedded in sand on a raised platform some 12 or 13 inches high. The sides and back of the hearth consist of square prisms of iron, 6 to 8 inches thick, lying on each other and resting on the edge of the sump, forming a hearth some 16 to 18 inches deep, open in front. A sliding door of plate iron is sometimes provided.

[1] Rectangular blast furnaces.

A single horizontal tuyere enters at the back, a little above the sump. In front of the hearth is a sloping iron plate—the workstone—W. The upper edge of this is level with the sump, and the lower rests on the masonry platform. It measures about 3 feet by $1\frac{1}{2}$ feet, and has a raised rim running round the sides and lower edge, a groove being cut diagonally across it, from top to bottom, down which the metal runs after filling the sump. The lead-pot P is placed in front of the workstone.

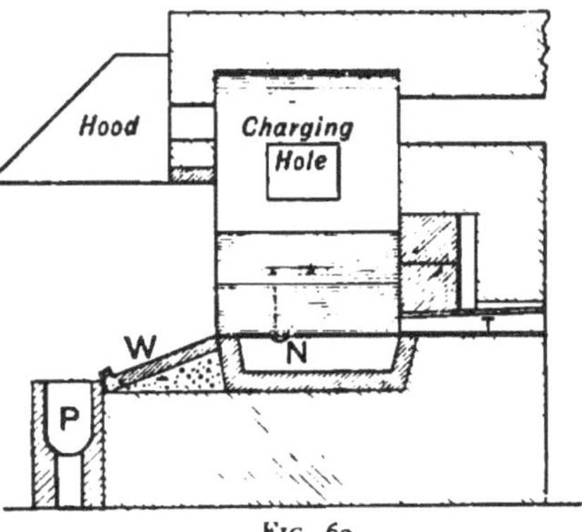

Fig. 62.

Note.—In some hearths the size above the workstone can be regulated by a movable cast-iron prism, which can be packed up to the desired height by fire-bricks and moved towards or from the back.

The charging hole is at the side.

Formerly raw ore was treated in these hearths, but it is now more general to partially roast and agglutinate the ore to prevent loss by being blown away and carried into the flues.

Coal and peat are used as fuel.

In Scotland the hearth is worked continuously, in shifts of 6 hours. In the north of England it works intermittently. The process is conducted as follows:—Assuming that the hearth is in operation, the sump being full of lead and the hearth a glowing red, a quantity of half-smelted material—browse—is thrown in *next the tuyere*, to assist in the distribution of the blast, and ore and fuel added. The hearth is kept full of materials. At intervals of a few minutes the workman draws the charge from the hearth on to the workstone, by means of a hooked bar, breaks up the glowing mass, and picks out the slags. The unsmelted portion is returned to the hearth, after an addition of lime, and fresh materials are added on top. Much lead drains out of the charge while on the workstone, and is conducted by the groove into the lead-pot, into which the reduced lead overflows from the sump.

The reduction of lead in this process is due, partly to "reaction," as in the Flintshire and analagous processes, and partly to direct reduction by the carbon of the fuel. The oxides and sulphate produced in the preliminary roasting, or *by the excess of air blown in*, if raw galena is being smelted, react on the unaltered portions of the sulphide, while the oxide is reduced to some extent by the fuel. No desulphurizing agent is added, as in the water-jacket furnace and Cornish process.

The addition of lime stiffens the charge. If the slag melts too easily, excess of silicate of lead is present, which, by its ready fusion, may enclose portions of the charge and prevent its reduction, in addition to the loss of lead in the slag. The slag consists of silicates of lead and lime, with sulphate and sulphide of lead, and other bodies. It is smelted in the usual way.

A hearth yields about 70 cwts. of lead in 24 hours, and consumes about 12 cwts. of coal.

The metal from the ore hearth is of good quality, owing to the operation being conducted at too low a temperature to effect the reduction of the impurities. The loss of lead in the slags is small, being less than 4 per cent. of the lead in the ore. The loss as fume is much greater with raw than with roasted ore. It varies from 7 to 20 per cent. of the lead.

A blind chamber behind the hearth is provided for the deposit of portions of the charge carried off by the violence of the blast. These are known as "hearth ends."

Softening or Improvement of Hard Lead.—The pig lead, as obtained from many operations, contains various impurities, antimony, tin, copper, zinc, sulphur, iron, silver, and bismuth being often present. Their effect is to harden the lead and unfit it for the purposes to which it is generally applied. The foreign metals are removed by oxidation, the lead being exposed at a red heat to the action of the air in a reverberatory furnace, the bed of which sometimes consists of a cast-iron pan—some 10 feet long, $5\frac{1}{2}$ feet wide, and 10 inches deep—or one of wrought iron, with a fire-brick lining, or is made of slags. In the latter cases higher temperatures can be employed, and the process occupies less time. The lead is ladled or run in from a melting-pot, or the pig lead may be melted in the furnace.

The oxides which form, consisting of oxide of lead mixed with those of the impurities, are skimmed off from time to time, to expose fresh surfaces, lime being added, if fused, to stiffen them. Samples of the metal are withdrawn and cast. When the lead shows a peculiar flaky appearance, the operation is judged complete, and the lead is ladled out or run into iron pig moulds.

The oxides which form in the earlier stage of the refining are richer in tin; those which form later on, in antimony.

When much copper is present, the lead is *liquated* before softening, as that element is not removed to any great extent. This is accomplished at Clausthal on the bed of a reverberatory furnace, which slopes slightly upwards from the fireplace. The temperature of the flue-end of the hearth is below the melting-point of lead. The metal is introduced there, and gradually moved forward. The lead melts and drains away, and the residues are moved gradually nearer the fire to sweat out all the lead, and are then raked out of the furnace. This residue contains the copper and nickel and cobalt, and often some arsenic and sulphur.

Reduction of Litharge and Drosses.—The "drosses," or "abstriches"—impure litharge—formed during softening or other operations, are reduced by intimately mixing them with small coal, grinding them together under edge-runners, and smelting in a reverberatory furnace, the bed of which is protected from corrosion by a layer of coke; this is formed by introducing a few inches of moistened caking-coal slack into the furnace, and beating it down. The bed slopes slightly, and the reduced lead drains into a basin in front.

Small water-jacketed furnaces are now largely employed for this purpose. The hard lead obtained, marked H, is richer in antimony, etc., and is again softened. The dross yields on reduction hard hard—HH—lead, and so on. This is repeated, until lead rich in antimony—sometimes more than 50 per cent.—is obtained. This is sold to the type-founders (see Alloys).

Desilverization of Lead.—Silver, as before noted, is commonly found in lead ores, and during smelting passes into the metal. Lead containing more than 3 ounces to the ton is treated for its extraction. Two methods are followed for this purpose.

Pattinsonising—In Pattinson's process, advantage is taken of the fact that alloys of lead with silver containing less than 1·8 per cent.[1] of silver, have a lower melting-point than pure lead, and that lead in the solid state is denser than when molten. In consequence of this, if a large body of lead be melted and cooled slowly with constant stirring, the lead which first crystallizes is poorer in silver than that which remains fluid. By removing the crystals with perforated ladles, a fluid alloy richer in silver and a lead poorer than the original are

Fig. 63.

obtained. The crystals removed are, of course, covered with the fluid alloy, and thus, in the process of removal, some silver is also carried off. By repeating the process on the enriched alloy, the silver contents of the fluid portion are again increased, until an alloy containing sufficient silver to be cupelled is obtained. Or the rich alloy may be treated by the Parke's process, which see.

[1] 640 ozs. per ton.

The process is conducted in a series of iron pans set side by side, as shown in Fig. 63, each capable of holding from 10 to 15 tons of lead. A 15-ton boiler is 5 feet 2 inches in diameter, and has a capacity of 43 cubic feet. A full set of pots numbers 13. Each pot is heated by its own fire, which is controlled by a damper. The products of combustion pass from the fireplace into a flue encircling the pot, and thence to the main flue.

The crystals are removed by a perforated ladle made of half-inch iron plate, 16 to 20 inches in diameter, 4 to 6 inches deep, with a handle about 9 feet long, about half the length being iron and the other wood. A chisel-ended iron bar is used to break up the lead crusts and for stirring, and a flat, perforated shovel for skimming.

Small pots filled with melted lead, between each pair of boilers, are sometimes provided for keeping the ladles hot, and the range of boilers is commanded by a crane for the transference of the ladlesful of crystals removed, or pivoted rests some 18 inches high with a roller top are placed between the pots. These are used as fulcrums for resting the handle of the ladle while fishing out the crystals and "turning over" from one pot to the next.

The lead to be desilverized is first melted in one of the pots, and got sufficiently hot to oxidize. A scum of dross forms, and is removed. (If very impure, it is necessary to liquate and soften before Pattinsonizing.) The fires are then withdrawn, and the cooling down facilitated by sprinkling water on the surface. The crusts of lead which form are pushed down into the metal, until there is difficulty in melting them, when the water is stopped and the bath thoroughly paddled. As the mass further cools, the lead begins to solidify in crystals. These are larger as the alloy is poorer in silver. Being heavier than the molten alloy, they sink, and the lead requires to be continually stirred and broken up, to prevent the formation of masses of crystals, which would entangle rich lead. The temperature must be carefully managed, or the crystallization will either be too slow, or masses of crystals will form. When the formation of crystals has progressed sufficiently, they are removed by the ladles, and transferred to the next pot to the left, which is already hot enough to melt them.

Each ladleful as it is raised to the surface is allowed to drain, and shaken to remove the liquid as much as possible. In this manner two-thirds or seven-eighths of the lead is removed. The former method is known as the "high" and the latter as the "low" system.

Working on the high system all the lead removed is thrown into the next pot. In the "low" system of working the last eighth contains too much silver, and is thrown on the ground, to be used with lead of the same richness in silver. The fluid alloy remaining in the pot is transferred to the next pot to the right. Working on the high system, it contains about twice the proportion of silver in the original lead, while the poor lead passing to the left averages about one-half.

By repeating the process on the enriched lead the proportion is again doubled, while the poor lead, similarly treated, will be again halved. Starting with a 10-oz. lead, one-third rich lead assaying 20 ozs., and two-thirds poor lead assaying 5 ozs. to the ton will be first obtained. On again treating the enriched alloy, 40 ozs. rich and 10 ozs. poor will be obtained. A third treatment yields 80 ozs. rich and 20 ozs. poor; a fourth, 160 rich and 40 ozs. poor; a fifth, 320 ozs. rich and 80 ozs. poor, and so on.

The *poor* lead, from the first crystallization, on a second treatment, yields 10 ozs. rich and $2\frac{1}{2}$ poor; a third treatment gives 5 ozs. rich and $1\frac{1}{4}$ poor; and a fourth treatment gives $2\frac{1}{2}$ ozs. rich and $\frac{5}{8}$ poor. These figures are only approximate; in practice they are not always realized.

In actual work, alternate pots are generally being crystallized at the same time, so that the rich third from the one, and the poor two-thirds from the other, make up a charge for the intervening pot.

When less than 1 oz. of silver per ton is present, the poor crystals are transferred to the market pot on the extreme left, which has a capacity of only about two-thirds of the others, and from which the lead is cast into pigs.

The *rich* lead containing from 600 to 700 ozs. per ton is cupelled (see page 225).

The Pattinson process is now mainly followed for the

enrichment of leads too poor to be treated by zinc, as described (see Parkes's Process).

The oxidation of the lead which accompanies the repeated meltings so purifies it that, by the time it reaches the market pot, there is no need for further softening, and it is cast into pigs.

NOTE.—The copper, antimony, bismuth, and iron remain mostly in the fluid portion, and would give trouble—especially antimony—in the cupellation of the enriched lead. Before Pattinsonizing, it is "improved" if more than 0·5 per cent. impurity is present.

Rozan Process—Pattinsonizing by steam.—This method was introduced by Messrs. Luce and Rozan, at Marseilles, and has been adopted to some extent. The novelty consists in the method of stirring up the molten lead by means of steam at high pressure forced through the lead, the surface being cooled by water as before. The crystals are not removed, but the liquid enriched alloy is run off and the crystals remain in the pot. Lead of the same richness is run in from a melting-pot above, and the operation repeated. A great saving in labour, fuel, and drosses results.

Parkes's Process—Desilverizing by zinc. This process has displaced, to a very large extent, the Pattinson method of desilvering. Or the "work lead" is Pattinsonized until it has attained a richness of about 40 to 60 ozs., and is then treated with zinc.

The process depends on two facts. *First*, that zinc and lead melted together do not alloy, but separate according to their specific gravities, the zinc rising to the surface, carrying only some 2 per cent. of lead.

Second.—Silver (as well as copper, etc.) alloys with zinc more readily than with lead, and hence, when that metal is mixed with argentiferous lead, the silver is collected in the zinc scum which rises to the top and is removed.

The method of carrying out the process varies somewhat in different works, and the quantity of zinc required depends on the amount of silver present.

The arrangement of zincing portion of a Parkes plant is shown in Fig. 64. The two large pots, A, are capable of holding about 25 tons of lead, and are employed in the addition of the zinc. The smaller ones, B, have a capacity of about 6 tons, and are for the treatment of the zinc crusts on removal. D is a reverberatory furnace for removing the zinc taken up by the lead, by oxidation.

The lead is melted in one of the pots A, heated to the melting-point of zinc, and skimmed. A portion of the zinc is then added, and when this has melted, a further addition of zinc is made, and the whole is well paddled for some 15 minutes. The pot is then covered over and allowed to rest for a period, varying from 1 to 3 hours.

While at rest, the zinc gradually rises to the top, carrying with it the silver. As it cools, it forms a crust on top, in which a good deal of lead is entangled. The zinc crust is removed by a ladle into the middle of the smaller pots, and the pot thoroughly skimmed till the lead begins to set. The

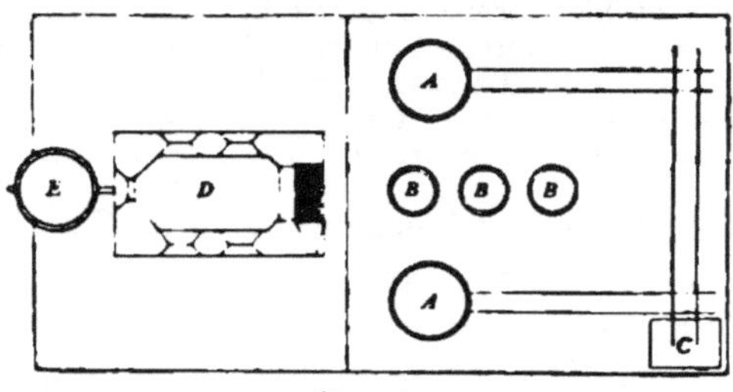

Fig. 64.

pot is again heated up, and a second addition of zinc made to the lead, well paddled in, and again allowed to cool. The amount added depends on the amount of silver remaining in the lead. The crusts formed are removed as before. After this second treatment, the lead is desilverized and is then run or syphoned off into the improving furnace D, to remove the zinc present in the lead. This amounts to about $\frac{1}{2}$ per cent. The lead is skimmed from time to time. Samples taken out are cast in moulds and examined. When the surface indicates a sufficient degree of purity, the lead is run from the furnace into a lead-pot, E, allowed to cool down, and cast.

Note.—With lead containing more than 80 ozs. of silver, the addition of the zinc in three portions is advisable.

The first crusts removed to the smaller pots are gently heated to liquate out the adherent lead. This is either cupelled, or generally returned to the zincing-pot with the next charge. After liquation, the crusts are transferred into the right-hand pot, and sent off to be distilled (see p. 197). The last crusts are used as the first addition of zinc made to the next charge.

The total amount of zinc required varies. A 20-oz. lead requires 30 lbs. of zinc per ton, equal to 1·33 per cent.; a 40-oz. lead requires 35 lbs., equal to 1·56 per cent.; and a 60-oz. lead about 38 lbs., equal to 1·69 per cent.; while a 500-oz. lead requires about $2\frac{1}{2}$ per cent.

In **Cordurie's** process, the zinc to be added is enclosed in a perforated cast-iron box, fixed on the end of a revolving vertical shaft. Immediately above the box is a propeller-shaped paddle, which as the zinc melts distributes it uniformly through the lead and thoroughly mixes it. Three zincifications are made.

The softening is effected by running the lead into a pot situated at a lower level, heating it to redness, and blowing superheated steam through it, succeeded by a mixture of steam and air. The zinc and iron present decompose the steam and are oxidized, hydrogen being liberated. Later, the copper and antimony remaining are oxidized by the air.

Treatment of Zinc Crusts.—The zinc crusts carry in addition to the silver a large proportion of lead, together with the copper, and some antimony, arsenic, and nickel.

The zinc is distilled off in large plumbago crucibles, provided with condensers, as shown in Fig. 65, about 18 inches in diameter and 27 to 30 inches high, provided with a cover, as shown. A clay pipe leads from a hole in the side to the condenser, C, standing in front, in which the recovered zinc condenses. A little lime and coal-dust are often added. The residual lead is cast in moulds, and afterwards cupelled. Any bismuth, antimony, copper, etc., in the crusts remain with the lead.

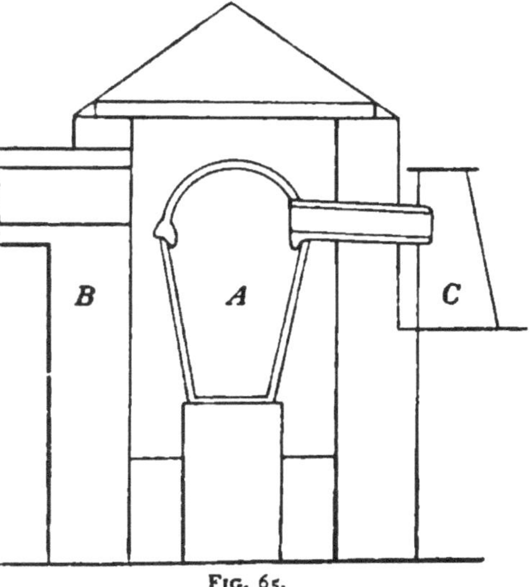

Fig. 65.

Lead Fume.—The gases passing away from the various furnaces carry with them considerable quantities of dust and volatile compounds of lead. These are deposited in the flues

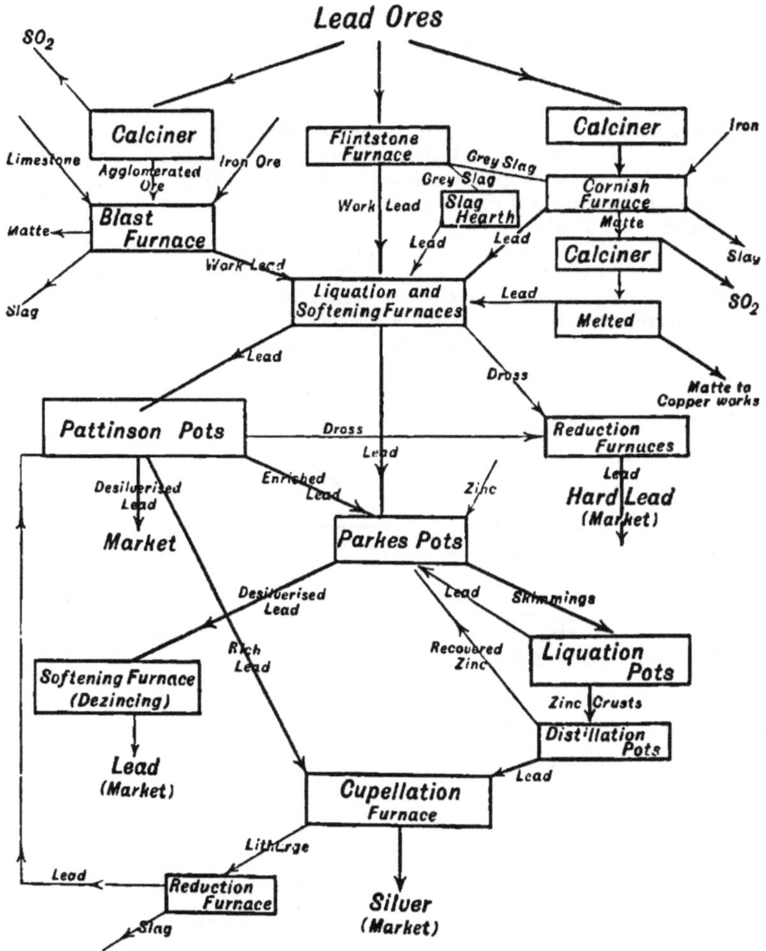

FIG. 66.—Synopsis of Lead Smelting and Desilverizing Processes.

through which the gases are conveyed to the chimney-stack, and constitute "lead fume." It consists of lead oxide, sulphide, and sulphate, with some lime, oxide of iron, alumina, etc., carried over as fine dust, and some silver. Zinc oxide is often present, especially in blast-furnace flues. A crust of this substance is often found in the upper part of the furnace when smelting zinciferous ores in blast furnaces.

For the condensation of the fume, and to minimize the nuisance arising from noxious vapours, the flues are sometimes from 3 to 5 miles long, and measure as much as 8 feet by 9, with which all the furnaces are connected.

Staggs, Stokoe, and French and Wilson have introduced condensers in which the gases are washed, or caused to pass through a wet filter of faggots or gauze, and thus deposit the solid matter.

The fume is smelted in blast furnaces, being previously agglomerated by heat to facilitate charging.

CHAPTER XIV.

MERCURY.

THIS is the only metal which is fluid at ordinary temperatures. About $-39°$ C. it freezes to a leaden grey, hard, malleable mass, contracting considerably on solidifying.

Its silvery white colour, and the readiness with which it moves about, owing to not wetting the surfaces (except metallic) with which it comes in contact, has earned for it the name of quicksilver (German, "quick silber"). Its specific gravity is about 13·6, which when solid becomes about 14·2. At a temperature of 357·25° C. it *boils*, giving a colourless vapour, but gives off vapour at much lower temperature, even below 40° C.[1] (see Condensation). Its low specific heat, high conductivity, and high boiling-point render it suitable for the construction of thermometers; its fluidity and high specific gravity for barometers. When in a very fine state of division,

[1] A gold leaf suspended over mercury at the ordinary temperature is whitened in course of time, being attacked by the vapour evolved.

and a film of foreign matter interposes, the globules will not unite together, the mercury is said to be *floured.*

It is permanent in air, oxygen, etc., at ordinary temperatures, but when heated near its boiling-point in air it oxidizes, forming red oxide of mercury. This is reduced to metal, with the separation of oxygen at a somewhat higher temperature. It is attacked by chlorine and ferric and cupric chlorides.

It is unattacked by hydrochloric acid, and but slowly, by sulphuric acid, unless hot and concentrated, when sulphurous acid gas is evolved and sulphate of mercury formed. Strong nitric acid rapidly dissolves the metal, but when dilute and cold has little action. Mercury compounds are readily decomposed by iron, copper, and other metals.

Mercury combines directly with sulphur, producing mercuric sulphide, or vermilion. It is prepared by heating mercury and sulphur together in an iron pan, with constant stirring, when a black brittle mass is produced. This is introduced at intervals into long, upright, iron retorts, or tall earthen jars, the lower parts of which are heated to redness. The sulphide volatilizes and condenses in a crystalline form in the cool upper parts. This deposit is *red*, and is ground, levigated, and dried. It is the *vermilion* of commerce.

Amalgams.—Mercury attacks and dissolves most metals, forming liquid alloys when the mercury is in excess. If the excess is removed by squeezing through chamois leather, a semi-solid amalgam is generally obtained. The remainder of the mercury is expelled on heating, and the other metal remains.

Gold, silver, zinc, tin, lead, antimony, bismuth, copper, and the alkali metals may be amalgamated by addition to mercury. Copper is best amalgamated by decomposing a salt of mercury by metallic copper, as the surface is not readily attacked by the metal. Mercurous nitrate is generally employed. Iron is not attacked directly, but iron amalgam may be obtained by the electrolysis of ferrous chloride with a negative pole of mercury.

The presence of these amalgams renders mercury less mobile, and when base metals are present, the oxidation which takes place, owing to the fine state of division of the metal in solution, causes the mercury to leave a "tail" behind it, if run down a slightly inclined porcelain tile. When pure it leaves no tail.

The amalgam with tin is used for silvering looking-glasses; amalgams with copper, tin and cadmium, silver and gold are used as tooth stoppings. The density of the copper amalgam is the same solid as plastic, to which state it may be reduced by slightly warming and working in a mortar. It is used for sealing bottles.

Metals are not readily attacked by mercury unless the surface is clean. Hence the presence of free acid aids amalgamation by removing films of oxide, etc. Sodium amalgam is often added to mercury in the amalgamation of gold and silver ores to prevent the mercury becoming "dead" and inactive by the oxidation of other metals, such as copper, etc., which may be taken up by it. Such mercury is apt to get into a finely divided state, the oxide films preventing the globules from coalescing, and it "sickens" or becomes "floured," in which case both the mercury and the precious metal it contains will probably pass into the residues or "tailings" and be lost. The sodium, by liberating hydrogen from the water present at the surface of the globules, prevents oxidation.

The silvering of mirrors is accomplished by squeezing mercury from a chamois leather bag over a sheet of tinfoil lying on a polished slab, forming a thin film of the amalgam. The carefully cleaned glass is then pushed gradually on, taking care to prevent air bubbles getting between, covered with felt and weighted. By inclining the slab and increasing the inclination from time to time, the excess of mercury is drained away and the amalgam adheres to the glass. The resulting film contains about 20 per cent. of mercury and 80 per cent. of tin.

ORES OF MERCURY.

"Native" Mercury occurs in globules in cinnabar, and amalgams of gold and silver are also found.

Cinnabar.—Mercuric sulphide (HgS) is the principal ore. It is a heavy mineral, of a vivid red colour; but some varieties are purplish. Its specific gravity is about 8. Large deposits occur at Almaden in Spain, Idria in Carniola, Bavaria, California, Chili, Peru, China, and elsewhere. Like hematite, it gives a red streak, but is volatilized by heat. The Idrian deposits have been worked about 400 years. Cinnabar when pure contains 85 per cent. of mercury, but the ores frequently contain less than 2 per cent., and are often bituminous in character. Mercury is often a constituent of fahl ore (p. 210).

Smelting, or Extraction.—The principles involved in the separation of mercury from cinnabar are very simple. When heated in a current of air, the sulphur is burnt off as SO_2, and the metal volatilized.

It therefore only remains to efficiently condense the vapour. This, owing to the readiness with which the metal gives off vapour, is a matter of much difficulty.

Cinnabar is decomposed when heated with lime, sulphide and sulphate of lime being produced thus :—

$$4HgS + 4CaO = 3CaS + CaSO_4 + 4Hg$$

Iron reduces it to mercury ; sulphide of iron resulting.

Idrian Furnace.—Fig. 67 shows the furnace employed at Idria. The cinnabar is placed on arches *n*, *p*, *r*, in the central chamber, over the fireplace. The larger lumps are placed in the lowest arch. The upper arch is occupied by small ore or

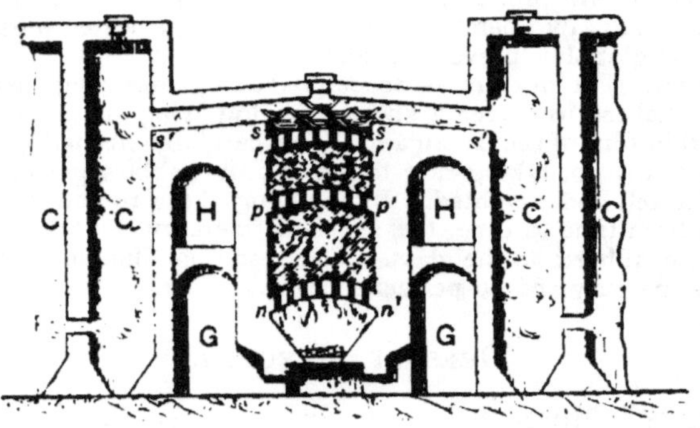

Fig. 67.

dust, on trays, as shown, or moulded into blocks with a little clay. The products of the combustion and the SO_2 and mercury vapour are led to the condensing chambers C, of which there are six on each side, by the passages *s'*. Each chamber communicates with the next alternately at top and bottom. The greater part of the metal is condensed in the first 2 or 3 chambers. The remainder is deposited as soot or dust in the succeeding chambers. The floors of these chambers incline towards an outlet at the side, by which the condensed mercury drains away, and is carried by a channel to the locked tank in which it collects. In the last chamber the condensation is assisted by water spray or by canvas screens stretched across it, covered with wet sawdust. The furnace and condensers are about 180 feet long and 30 feet high. The charge for the double furnace is nearly 100 tons. The operation takes about a week to complete, of which five days are occupied in cooling, and only about 12 hours in distillation. About 4 tons of mercury are obtained from each charge.

In Hahner's modification of the Idrian furnace, the ore, mixed with charcoal, is fed into the central shaft from a hopper above, and the furnace works continuously. The condensing chambers are prevented from becoming overheated by covering them with iron plates and cooling by a spray of cold water. The spent ore is removed at intervals through the grate at the bottom of the shaft.

Aludel Furnace.—Fig. 68 shows the Aludel furnace in use at Almaden in Spain. The ore is placed in the chamber F,

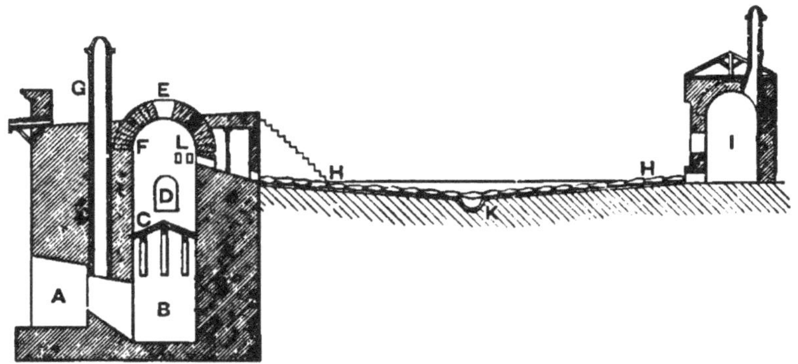

FIG. 68.—B, fireplace; C, perforated arch; F, ore chamber; D E, charging openings; G, chimney for fire; L, openings to aludels; H, aludels; K, mercury gutter; I, condensing chamber.

resting on perforated arch C, over the fireplace B. A quantity of spent ore or quartz is placed in the bottom, then poor ores followed by richer ones. The powdered ores are made into balls and placed on the top. A wood fire is first made in B, and the whole thoroughly heated. The fires are then withdrawn and air admitted. In passing through the grate, spent ore, etc., at the bottom, it becomes heated, and calcines and reduces the cinnabar. The vapour and gases pass out by the passages L, and through rows of "aludels" resting on sloping masonry roofs or benches. The aludels are earthenware, pear-shaped condensers (Fig. 69) 16 inches long; the neck is $4\frac{1}{2}$ inches, the wider end about 7 inches, and the middle 11 inches in diameter. These are fitted together and luted with clay. The middle ones have a hole on the under side which permits the condensed mercury to drain into the trough K, by which it is conveyed away. From the

Fig. 69.

aludels the vapours pass into the chamber I, from which they escape by the small chimney. The operation lasts about 24 hours, and the cooling down 3 or 4 days. The condensation in both these furnaces is imperfect.

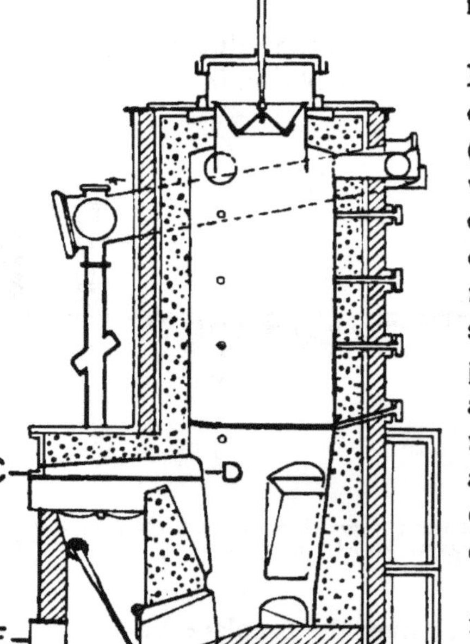

Fig. 70.—Californian Furnace.

Muffle, or Retort, Furnaces are used for reducing the pure "fines" (small ore), and the fume which collects in the condensers nearest the ore chamber, which consists mainly of sulphide and sulphate. From 10 to 20 per cent. of quicklime is added and the mixture made into bricks. These are heated and the vapour condensed in iron tubes dipping under water.

The Alberti Furnace is a long-bedded, reverberatory furnace, the flues of which consist of large water-cooled iron pipes. Poor ores are treated in these furnaces, but the acid vapours attack the iron.

Channel Furnaces.—The beds of these furnaces are steeply inclined planes divided into channels down which the ores rickle, being roasted by an ascending current of air and hot gases from a fire situated at the bottom of the incline. The vapours are led into condensers.

Shaft Furnaces are employed in California. These furnaces work continuously. The ore chamber D (Fig. 70) is cylindrical in form, standing on a hexagonal base. Three fireplaces C, with ash-pits, etc., communicate with the chamber on alternate sides of the hexagon. Below the fireplaces the chamber contracts and the calcined ore is withdrawn through openings at the side. The top of the chamber is closed by a cup-and-cone arrangement, the cup being covered by a gas-tight cover, which is always in position before the cone is lowered to admit the charge, thus preventing escape of vapours. The gases are led away by inclined iron pipes to condensers. Peep-holes, for the inspection of the furnace, are also provided. The chamber is 19 feet from base of cone to bell, and 6 feet wide. It roasts about 10 tons per day. In starting, the shaft is filled to the level of the fireplaces with spent ore, and then to within 3 feet of the top with ore mixed with 1 to 2 per cent. of coke or charcoal. The fires in C are lighted, wood being used as fuel, and the whole furnace heated to full redness. Some spent ore is then removed through E and fresh ore admitted from the top. Fresh additions are made about every two hours.

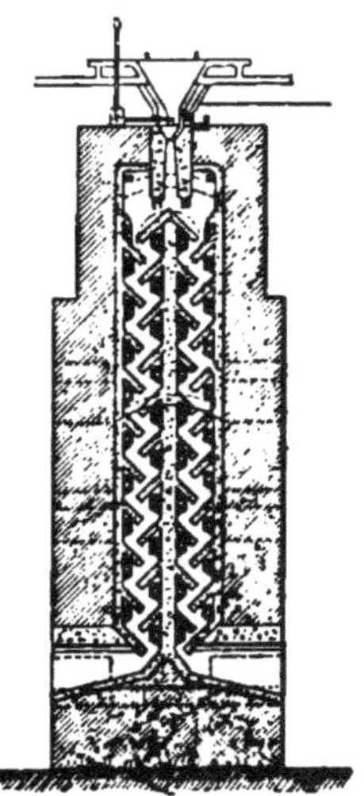

Fig. 71.

Hüttner and Scott's furnace for the continuous treatment of fines is shown in Fig. 71. The fine ore is fed in from the hopper at the top of the chamber, on to a series of sloping shelves, and passes in a zigzag fashion down the furnace, being turned over and over in its descent. The chamber is 27 feet high, $25\frac{1}{2}$ inches wide, and 11 feet 6 inches long. At one end is a fireplace supplied with hot-air heated by iron pipes placed in the condensing chambers. The hot gases, mixed with hot air, are admitted to the ore chamber by a series of openings at one end, under each shelf, and pass away to the condensers by corresponding openings at the opposite end of the furnace. The furnace treats about 1½ tons per hour, and the spent ore is removed at intervals.

A continuous retort furnace erected by the "El Pouvenir" Company in Spain, is shown in Fig. 72.[1] The retorts A are of cast iron, and are supported above a fireplace B. The mouth inclines upwards and communicates with the condensing apparatus C, by a flue as shown. A hydraulic exhaust injector

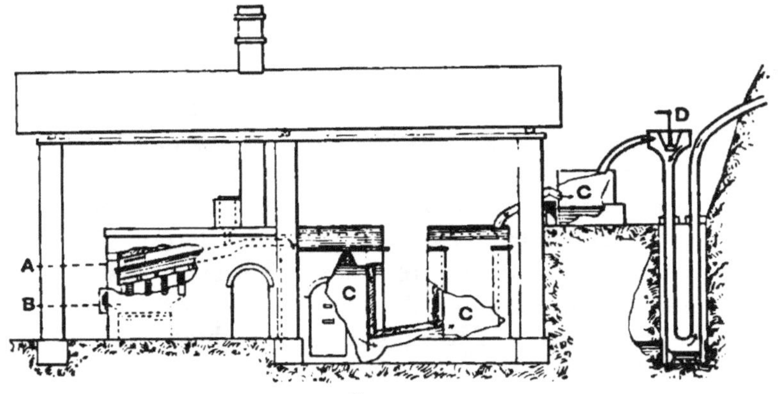

FIG. 72.

D draws the vapours through the condenser and permits of the lower end of the retort being opened to remove a portion of the charge without escape of vapour. The ore is fed in at the top, about ½ cwt. every hour and a half, giving an average of ¾ ton per retort per day. Rich ores are mixed with lime. Two large condensing chambers are provided. From the second they pass into a smaller chamber containing water, and then to the exhaust.

In shaft furnaces at the same works the hydraulic exhaust is also employed. The fireplace is placed below a perforated arch, as at Almaden, but the top is provided with charging apparatus, and the ore is discharged after calcination through openings at the side.

The condensation of mercury offers a difficult problem. In calcining furnaces of all types, the large volume of gases to be cooled (products of

[1] "Journal Soc. Chem., Ind.": 1890, p. 93.

combustion of fuel, nitrogen of the air, sulphur dioxide, and mercury vapour), and the ease with which mercury gives off vapour, render its perfect recovery difficult. The gases often contain less than 1 per cent. by volume of mercury vapour. Except in the condensers nearest the ore chambers, iron cannot be used on account of the acid liquors (H_2SO_3 and H_2SO_4), which condense when sufficiently cool. Other metals cannot be used as they are attacked by mercury.

It is therefore necessary to employ large condensers, as at Idria, to slow down the current to completely cool them, and to admit the vapour as near the boiling-point as possible. With furnaces working continuously, auxiliary cooling appliances, such as earthenware or iron pipes, with or without water cooling, are necessary. Glass chambers in wooden frames have been largely adopted for gases comparatively cool, communicating alternately at the top and bottom. Mercurial vapours are highly poisonous, producing salivation.

Purification of Mercury.—Commercial mercury often contains lead, zinc, bismuth, and other impurities. The presence of these may be detected by allowing it to run down a white tile. If impurities are present, the metal leaves a tail. Mercury is purified by squeezing through chamois leather and subsequent re-distillation. It may also be purified by exposing it in a thin layer to the action of dilute nitric acid, mercurous nitrate, or ferric chloride solution. The impurities are dissolved together with some mercury. It comes into the market in screw-necked iron bottles, containing $\frac{1}{2}$ to $\frac{3}{4}$ cwt. each.

CHAPTER XV.

SILVER.

Physical Properties.—This metal is characterized by its whiteness and brilliant lustre. It is somewhat softer than copper, but harder than gold. It is exceedingly malleable, being in this respect only inferior to gold, with which it may be alloyed without seriously impairing the malleability of that metal. It is highly ductile, and has a tensile strength of 14 tons per square inch. Its specific gravity is 10·5. It is the best conductor of heat and electricity. At about 950° C. it melts, and at high temperatures is sensibly volatile. In the electric furnace it boils and distils.

Chemical Properties.—The metal is unoxidized when

heated in air or oxygen, but molten silver *dissolves* about 22 times its volume of oxygen, which is given out on solidifying, the metal often being projected from the surface in curious growths. This phenomenon is known as "spitting," and does not occur if the metal is impure. The metal contracts on cooling. Silver oxide otherwise produced is decomposed by heat into silver and oxygen.

Silver combines readily with sulphur, forming silver sulphide (Ag_2S), a soft, dark-grey, fusible body. The blackening of silver when exposed, is due to the formation of this body by sulphur compounds in the atmosphere. This compound is also precipitated by adding sodium sulphide to a soluble salt of silver.

Sulphide of silver roasted in air is partly decomposed, sulphur dioxide and silver resulting, and is partially converted into sulphate. This conversion into sulphate takes place to a greater extent in the presence of sulphides and sulphates of other metals. Silver sulphate is soluble in water containing free sulphuric acid, and is decomposed by heat, metallic silver resulting. Silver sulphide is converted into chloride by the action of ferric, cuprous, and cupric chlorides.

Silver combines directly with chlorine, forming silver chloride, which is not decomposed by heat. This compound is also produced when hydrochloric acid, or a soluble chloride is added to a silver solution, and by roasting the sulphide with salt in the presence of moist air. It is insoluble in acids,[1] but dissolves in strong brine (sodium chloride) and other chlorides (especially ferric and cupric chlorides), in sodium thiosulphate (forming $Ag_2S_2O_3.Na_2S_2O_3$, if the sodium salt is in excess), in potassium cyanide (forming AgCN.KCN), and in ammonia. It fuses at a red heat, and is volatile at high temperatures.

Chloride of silver is reduced by "nascent" hydrogen, mercury, and by fusion with carbonate of soda.

$$AgCl + H = Ag + HCl$$
$$2AgCl + Na_2CO_3 = 2NaCl + CO_2 + O + Ag_2$$

[1] AgCl is somewhat soluble in hydrochloric acid. The strong acid dissolves 1 in 200 parts, and when diluted with an equal bulk of water, 1 part in 600.

Silver is deposited from solution in the metallic state by zinc, copper, iron, and other metals, and cuprous oxide.

Sulphuric acid dissolves it when heated, forming sulphate.

$$Ag_2 + 2H_2SO_4 = Ag_2SO_4 + 2H_2O + SO_2$$

Nitric dissolves it readily, silver nitrate being formed.

$$6Ag + 8HNO_3 = 6AgNO_3 + 2NO + 4H_2O$$

Hydrochloric acid has no action upon it.

Silver nitrate ($AgNO_3$) is a white solid, soluble in water. It crystallizes in flat tabular crystals. It is fusible without decomposition, but at a higher temperature, much below redness, it gives off oxygen and forms $AgNO_2$. At a red heat it is decomposed, yielding metallic silver.

This is made use of in the separation of silver and copper nitrates. The latter decomposes at a much lower temperature than the silver nitrate, and by careful heating may be resolved into oxide, leaving the silver nitrate unaltered. When a sample, treated with water, gives no blue colour on the addition of ammonia, the mass is boiled with water, to dissolve the silver nitrate, and filtered from the insoluble copper oxide. Boiling the mixed nitrates with silver oxide also throws down the copper as oxide.

Large quantities of nitrate of silver are produced in parting silver and gold.

Alloys.—Silver is too soft for use in the pure state, and is hardened by alloying it with copper. English sterling silver "standard silver" consists of an alloy of 925 parts of silver per 1000 alloyed with 75 of copper. This is equivalent to 11 ozs. 2 dwts. of silver per lb. *troy*. Alloys which contain more silver per lb. are described as "better," and those containing less as "worse" than standard. The Indian rupee, 11 ozs. 8 dwts. per lb., is 6 dwts. better, and the French standard alloy contains 10 ozs. 16 dwts. only, and is described as 6 dwts. worse.

The degree of purity is now often expressed in parts of silver per 1000; thus "900 fine" implies that the alloy contains 900 parts of fine silver and 100 of alloy per 1000.

Frosted Silver.—Silver is frosted by heating silver alloyed with copper in air. The *copper* oxidizes, and the oxide is dissolved off with sulphuric acid or ammonia, or by boiling with cream of tartar and salt. This leaves a "dead" surface consisting of finely divided silver.

Oxidized Silver.—This effect is produced by treating the silver with a soluble sulphide, such as sulphide of potash, and is due to the films of silver sulphide formed.

Ores of Silver.

"**Native**" **silver** occurs associated with ores of the metal; with gold, in electrum; with mercury, in amalgam.

Silver Sulphide (Ag_2S)—Argentite—occurs as a soft, malleable, grayish-black substance, which is readily fusible. It contains 87 per cent. of silver. Deposits containing it in a state of purity occur in Norway, Hungary, Saxony, Bohemia, Mexico, and the United States. It is the principal ore of silver.

Horn Silver—Silver Chloride ($AgCl$)—is found in South America. The bromide and iodide also occur.

Pyrargyrite.—Dark-red silver ore is a sulphantimonide of silver ($3Ag_2S.Sb_2S_3$) found in Mexico, South America, Transylvania, and elsewhere. **Proustite**—light-red silver ore—is a sulpharsenide ($3Ag_2S.As_2S_3$). **Stephanite** is another mineral of the same class.

Polybasite and **Argentiferous Fahl Ore** are compounds of copper, silver, arsenic, and antimony sulphides. The latter often contains other metals also.

Silver occurs in the ores of many other metals, probably as sulphide. Lead, zinc, and copper ores often contain it, and small quantities occur in iron pyrites and mispickel (arsenical iron pyrites). The production of silver from these sources is nearly one-half of the total extracted.

Extraction Processes.—The high price of silver permits of poor ores being treated and the adoption of more costly methods. Hence chemical methods, preceded by careful mechanical preparation, are often followed.

The treatment of silver ores proper may be divided into—

Amalgamation processes.
Wet processes.
Smelting with lead, or lead ores.
Smelting with copper ores.

Amalgamation Processes include those in which the silver is obtained as an amalgam with mercury, from which it is recovered by distillation and volatilizing the mercury. They may be divided into "floor," "barrel," and "pan" amalgamation processes. If not present as free silver or as chloride, the first step of the process is to convert the metal into chloride.

Floor Amalgamation.—In the "patio" process, still followed in Mexico and South America, the ores are hand-picked, and then contain some 80 ozs. of silver per ton, as native silver, chloride, and sulphide. Base ores containing large amounts of foreign sulphides are unsuitable for treatment by this process. The ore is first reduced to a fine state of division by stamping or grinding.

The *quimbalete* consists of a large boulder lashed to the middle of a long pole, rocking on a flat stone, worked by men sitting astride the ends of the pole, and working see-saw fashion, the ores being thrown under the boulder.

The *trapiche* is a large stone wheel, 6 feet in diameter and 5 feet across; the axle on which it revolves is attached to a perpendicular shaft driven by a horizontal water-wheel on the top. The wheel travels round a stone track, and the ores are gradually crushed.

The *ărastra*, for fine grinding, is a circular trough paved with hard stone. In the centre is an upright post to which projecting arms are attached. Heavy stones are lashed to these, by thongs of raw hide, and they are dragged round by mules attached to the ends of the arms, which project over the edge of the trough. Water is added, and, if much free silver or gold is present, a little mercury, to amalgamate them. The ore is thus reduced to mud.

The *Chilian mill* for grinding ores is in principle like an ordinary mortar-mill.

The operations are conducted as follows:—(1) The mud is taken to the amalgamating floor, or patio—a paved court—and spread out in a layer 6 inches to a foot thick. Some 3 to 5 per cent. of salt is added and well trodden in by mules for several hours, after which the heap is allowed to rest.

(2) Next morning a quantity of roasted copper pyrites—*magistral*[1]—is scattered over the heap, and some mercury. This, after mixing with shovels, is well trampled in; the turning over and trampling are repeated every other day for some days.

(3) Mercury to the extent of some 5 or 6 times the weight of silver present is sprinkled over the heap from canvas bags,

[1] This contains both copper and iron sulphates, and plays a material part in the reactions which occur.

and trampled in. If much antimony and arsenic, or other foreign sulphides are present, a hot solution of copper sulphate is added, together with copper precipitate (finely divided copper) (see p. 176), and thoroughly incorporated.

(4) After a further rest with repeated tramplings, a final addition of mercury is made to collect the amalgam, and after mixing, the stuff is conveyed to tanks, where it is stirred up with water, and the heavy amalgam settles out. The earthy matters are carried away by the water current.

The amalgam is treated in the ordinary manner (see p. 219).

In this process a complicated series of reactions occur. Copper chloride is formed by the reaction of the salt and copper sulphate.

$$CuSO_4 + 2NaCl = CuCl_2 + Na_2SO_4$$

This attacks the metallic silver, thus:

$$2CuCl_2 + Ag_2 = 2AgCl + Cu_2Cl_2$$

This cuprous chloride, which is soluble in the excess of salt employed, reacts on the sulphide of silver and converts it into chloride.

$$Ag_2S + Cu_2Cl_2 = 2AgCl + Cu_2S$$

Some free sulphur is also separated, probably thus:

$$Ag_2S + 2CuCl_2 = Cu_2Cl_2 + 2AgCl + S$$

The above reactions in some degree represent the chlorination, but the changes are very obscure. The silver chloride is decomposed by mercury, thus:

$$2AgCl + Hg_2 = Hg_2Cl_2 + Ag_2$$

the metal being dissolved by the excess of mercury. The operation occupies from 2 to 7 weeks.

NOTE.—The addition of copper precipitate is to ensure the reduction of the cupric salt to the cuprous state, or it will attack the mercury, forming calomel, and will thus increase the consumpt of mercury.

$$2CuCl_2 + Hg = Hg_2Cl_2 + Cu_2Cl_2$$
$$2CuCl_2 + Cu = 2Cu_2Cl_2$$

Formerly lime was added, to precipitate the copper, if in excess; but this hinders the chlorination by forming an inactive chloride.

Barrel Amalgamation was formerly practised at Freiberg.

The chlorination of the metal is effected by roasting the ore with salt as described (pp. 175 and 221). The roasted ore is next put into barrels, supported horizontally, capable of holding about a ton, and water added to make it into a stiff paste (pulp), some $1\frac{1}{4}$ to $1\frac{3}{4}$ cwt. of sheet-iron scrap added, and the barrels revolved on trunnions for several hours. The chloride is reduced by the iron, thus:

$$2AgCl + Fe = FeCl_2 + Ag_2$$

Mercury is then added to amalgamate the reduced silver, and the barrels again revolved some 16 hours. The contents of the barrels are then thinned by the addition of water, the amalgam collected together by slow revolution, and run off by a plug in the side. A little fresh mercury is added to collect any residual metal, and the barrels again revolved. This is run off as before. The residues are run into settlers and agitators—tanks with a current of water flowing through—by which the light matters are carried off and the amalgam (if any) sinks.

In the Krolinke, or Aaron process, in use at Benton, the roasting with salt is dispensed with. The chlorination of the silver is effected by an addition of cuprous chloride and salt. The cuprous chloride is prepared by boiling copper sulphate with salt and copper, or in other ways. The barrels may be arranged vertically or horizontally, and steam is blown in to heat the contents. Metallic copper is employed to reduce the silver, and the amalgamation with mercury takes place as before. The loss of mercury is greatly reduced, calomel not being formed. Aaron states that it can be brought as low as 2 lbs. per ton.[1] Iron borings are sometimes used for the reduction. Base ores can be treated by this process, a yield of 80 to 95 per cent. being obtained.

In both these processes there is considerable loss of mercury by "flouring"—that is, the mercury is broken up into fine particles, which will not coalesce to form a globule, and are carried off and lost. The addition of a little sodium amalgam is made to prevent this.

[1] *Iron*, Nos. 93 and 94.

Kettle Amalgamation (Cazo Process).—The ores treated by this process are mainly chlorides, bromides, and iodides. The ore is ground to a pulp in the mill, or arastra, and transferred to kettles with bottoms made of copper. From 5 to 10 per cent. of salt is added, and the mass heated with continual stirring. Mercury is added, and the heat is continued for some hours, till amalgamation is complete. The mass is then thinned with water, and the amalgam collected as before. The chloride, etc., is decomposed by the copper,

$$2AgCl + Cu_2 = Ag_2 + Cu_2Cl_2$$

yielding silver and cuprous chloride. This, in the presence of salt, reacts to some extent on the sulphides, after the

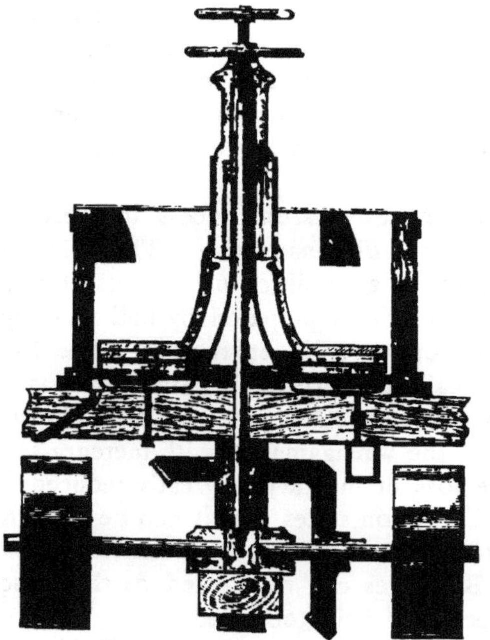

Fig. 73.—Amalgamatory Pan.

manner of the "patio" process, but sulphide ores generally retain enough silver to be subsequently treated on the "floor."

Pan Amalgamation.—The foregoing methods have generally given way to treatment in pans, a great saving in time being effected.

The pans employed in these processes vary somewhat in construction. One form is shown in Fig. 73. It consists of

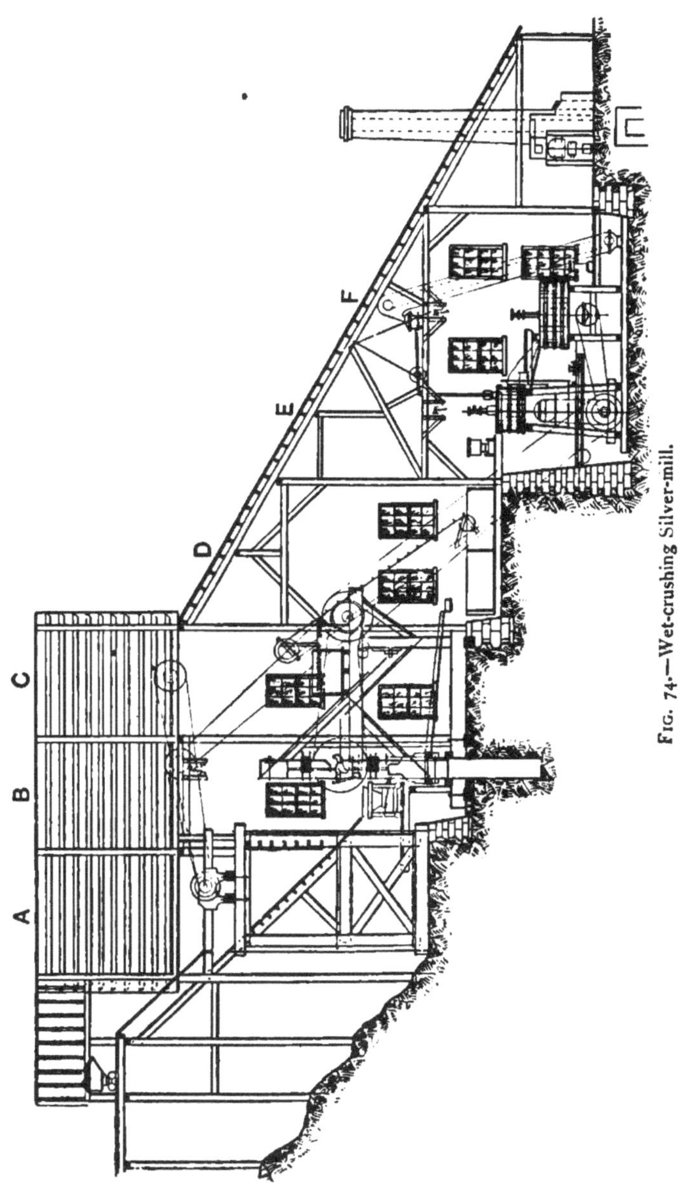

Fig. 74.—Wet-crushing Silver-mill.

a cast-iron pan some 5 feet in diameter, with a steam-jacketed bottom. Up the centre a hollow pillar rises, through which a shaft passes. To this the cast-iron muller is attached in a manner which permits of its being raised or lowered to any desired height by means of the hand-wheels on top. The crushed ore is ground between the flat faces of the muller and the bottom of the pan, motion being communicated by the bevel-wheel gearing under the bench on which the pan rests. Steam is admitted to keep the contents hot. A plug is provided for running off the pulp after amalgamation.

Instead of iron sides, wooden staves hooped with iron are employed, and copper bottoms and linings are sometimes employed.

Two methods of treating the ores are followed. In one, they are treated direct, and in the other they undergo a previous roasting with salt to chlorinate the silver.

In the *direct* process the ore is broken in a "stone-breaker," or "ore crusher" of the Blake type, A (Fig. 74), passes to a stamp battery, B, and is crushed "wet"—that is, with a supply of water (see Gold, p. 234), a 30-mesh screen being employed. From the battery the crushed ore passes over the amalgamated copper plates C to catch any free gold present, and then to the tanks D in which the mud settles.

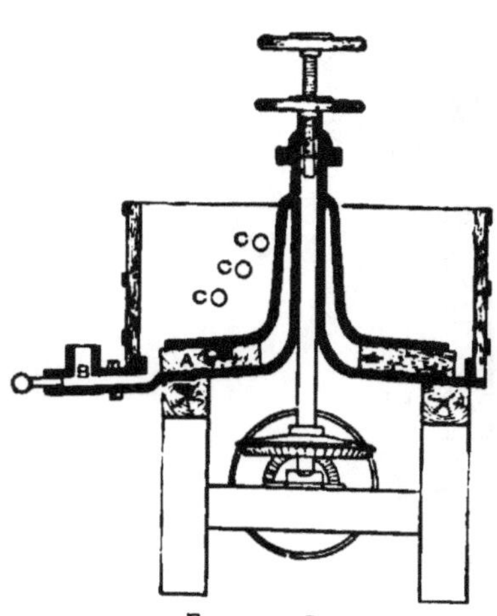

FIG. 75.—Settler.

The mud (pulp) is charged into the pans E, water added to a pasty consistency, and the muller lowered and revolved at the rate of 80 to 100 revolutions per minute. Salt and copper sulphate are also added, and the temperature is maintained at about 90° C. This grinding is continued some 3 or 4 hours. The pulp will then pass through an 80-mesh sieve. About 10 or 15 per cent. of mercury is then added, and the muller, somewhat raised, is again revolved for 2 or 3 hours, to thoroughly incorporate the mercury. The pulp is then thinned by the addition of water, and run off by the plug into the settler F, which resembles the amalgamator, save that the muller is replaced by a stirrer (Fig. 75), which makes some 10 revolutions per minute. Here the amalgam settles out. The mud is drawn off by opening the holes in succession, into another settler—the agitator—and then passed on to "frue vanners," or is treated on buddles, for the separation of the pyrites, etc. (concentrates), which often carry gold, while the light stuff is washed away.

In amalgamation processes preceded by roasting, the ores are crushed "dry."

In "dry" crushing, the ore, after being broken, is dried, a rotary furnace being employed. The dried ore is stamped, and the crushed ore falling through the screens is conveyed away by means of Archimedian screws, travelling belts, or elevators. Fig. 76 shows a "dry"-crushing mill. The powdered ore is next mixed with about 20 per cent. of salt, and roasted. Generally revolving furnaces of the Brückner type (Fig. 16) are employed. Stevelet calciners (Fig. 18), and long-bedded reverberatory furnaces are also employed. This roasting occupies about 8 hours. The ore is then transferred to the amalgamators and treated as before. The yield by this treatment is much greater than in wet crushing, but the items of labour and fuel consumption are increased, while the output of a plant is seriously diminished.

The loss of mercury is about 2 pounds per ton of ore treated. It is customary to add a little sodium or zinc amalgam, to keep the mercury from *flouring*, the hydrogen evolved keeping the mercury bright and *lively*, and preventing the formation of a film on the small globules, which prevents them from coalescing. Potassium cyanide, in small quantities, is often used for the same purpose. In wet crushing, mercury is also introduced into the mortar-box to retain gold.

In the roasting of *dry-crushed* ores with salt, there is a liability to form gold chloride, which is soluble, and will be lost if not completely decomposed in the pans.

In "wet" crushing, the sulphide of silver in the ore seems to be partially decomposed by the iron of the pan during amalgamation, with the formation of sulphide of iron, assisted by the cuprous chloride, produced when salt and copper sulphate are added.

The *best grinding* is secured with a thin pulp, and the *best amalgamation* with thick pulp, which prevents the mercury from settling out. It is usual to add *residues* to thicken the pulp prior to adding mercury. It is soft enough if the muller will turn in it.

Treatment of Amalgam.—The amalgam from the settlers

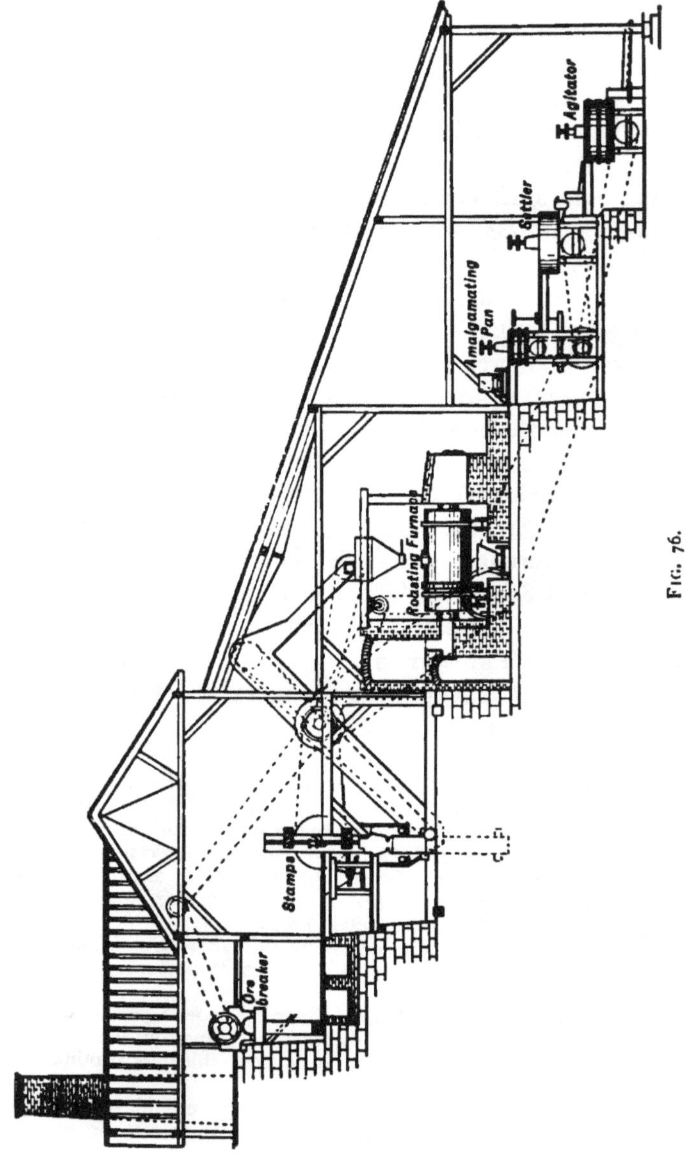

Fig. 76.

and agitator is often transferred to a smaller "clean-up" pan, and stirred with water to free it from heavy particles.

It is then strained through canvas bags, or squeezed through washleather, or by hydraulic pressure, in cylinders, the ends of which are made of wood cut across the grain. The excess of mercury which is thus removed is used again. It contains silver, but this is recovered in the subsequent working. The pasty amalgam which remains is then "retorted" to expel the mercury. One form of retort is shown in Fig. 78. The amalgam is put into the crucible, which is of iron, the head adapted, and the mercury as it distils off is condensed by the water-cooled tube. The crucible is coated with limewash.

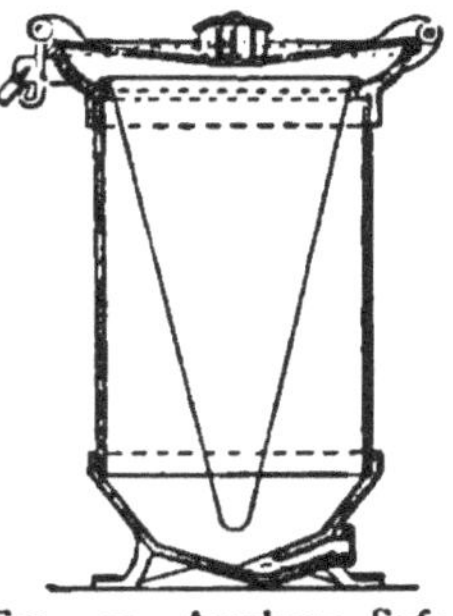

Fig. 77.—Amalgam Safe, with strainer.

The porous mass obtained is subsequently melted down in crucibles, and cast into bars weighing about 1000 ozs. The

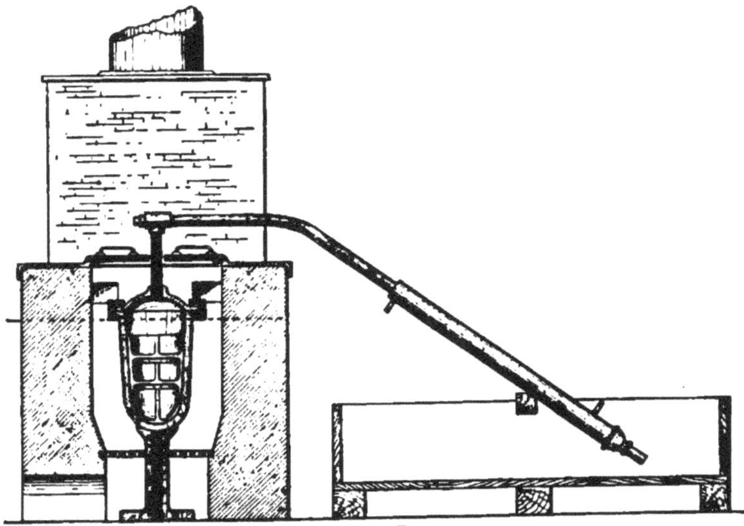

Fig. 78.—Retort.

crude bullion contains bismuth, antimony, copper, zinc, arsenic, etc. It is subsequently refined. This is partially effected by exposing the surface to the air while molten, and permitting the impurities to oxidize, the scum of oxides being scraped off. It is afterwards refined by cupellation.

Wet Processes.—Wet methods of silver extraction depend on the solubility of sulphate and chloride of silver in water or other solvent (see p. 208). They are largely employed in the treatment of copper mattes and bottoms, and certain classes of ore containing large quantities of foreign sulphides.

Ziervogel Process.—In this process the silver is converted into sulphate, by roasting. This is dissolved out by water, and the silver precipitated from solution by copper.

Sulphating roasting.—This depends on the fact already mentioned that if a mixture of iron, copper, silver, zinc, and lead sulphides be roasted, they are partly converted into oxides and partly into sulphates (see p. 173).

NOTE.—The production of sulphates is largely due to the formation of SO_3 by the SO_2 from the roasting mass, combining with oxygen under the influence of the "contact action" of the brickwork and siliceous matters in the ore—a sort of catalysis. This combines with the oxides. A *slow* current of air favours its production.

If carefully heated, the sulphates decompose in the order named. The SO_3, liberated in the decomposition of the iron and copper sulphates, attacks the silver, and tends to its complete conversion into sulphate.

Roasting Copper Mattes.—They are first roasted to remove the greater part of the sulphur, and then ground very fine and carefully roasted at a low and gradually increasing temperature in a double- or triple-bedded reverberatory calcining furnace. The matte is first introduced on the bed farthest from the fire, and is moved forward towards the fireplace. When the iron and copper sulphates formed during the roasting are nearly decomposed—determined by boiling a sample with water and observing the colour—the material is raked out.

Argentiferous copper ores are generally run down for matte, which is thus treated.

The roasted material is then treated with water containing a little free sulphuric acid—*leached*—in wooden tanks capable of holding about 1000 gallons. From these the liquor is run into settling tanks, at a lower level, and thence into tanks containing copper, where the silver is precipitated. Two sets of precipitating tanks are usually employed, the first containing heavy copper scrap or bars, and the latter precipitate and bean shot-copper.

$$Ag_2SO_4 + Cu = CuSO_4 + Ag_2$$

The copper is recovered by throwing it down with iron in similar tanks.

The residues contain the gold, a portion of the silver (owing to imperfect sulphating), copper, and iron as oxides, lead as sulphate. If bismuth and antimony are present in the matte, more silver is retained, owing to the formation of insoluble compounds.

The residues are smelted for copper by the "best select" process. The bottoms obtained are treated by electrolysis, or in a manner subsequently described.

Augustin's Process.—This consists of roasting the material mixed with salt for the purpose of converting the silver into chloride, which is then dissolved out by brine and precipitated by copper.

Chlorinating roasting.—The silver is converted—

(1) Into chloride either by the action of free chlorine, generated thus—

$$(a)\quad 2NaCl + O + SiO_2 = Na_2SiO_3 + Cl_2$$
$$(b)\quad 2HCl + O = H_2O + Cl_2$$

(2) Or by the action of hydrochloric acid gas, produced thus—

$$2NaCl + H_2O + SiO_2 = Na_2O.SiO_2 + 2HCl$$
$$2FeSO_4 + 4NaCl + 2H_2O + O = Fe_2O_3 + 2Na_2SO_4 + 4HCl$$

the moisture being present in the atmosphere of the furnace;

(3) By the action of chlorides of copper and iron produced by the reaction of sulphates on the salt added.

The leaching and precipitation are carried out as before.

In the treatment of copper bottoms, the two processes are often combined (at Freiberg, and some works in this country). The bottoms are granulated in water, and roasted to oxide, CuO, mixed with sulphur or sulphate of iron, and Ziervogelized.

The residues contain the gold and much silver, and are Augustinized. The gold passes into solution as chloride, and is, of course, precipitated by the copper. Great care is required in roasting, or the gold chloride will be decomposed by overheating, the metal remaining in the residue.

Claudet's Process is in extensive use for the recovery of silver from the cinders of iron pyrites used in vitriol manufacture, and is employed as an adjunct to the extraction of copper by Longmaid's process (p. 175). In the chlorinating roasting for copper, the silver is also chlorinated, and in the lixiviation with water is dissolved out by the excess of salt added in roasting. The first leachings, after cooling and settling in tanks—during which much lead sulphate and chloride separates out—are assayed for the amount of silver they contain. A soluble iodide is then added in sufficient quantity to precipitate it as insoluble silver iodide—

$$2AgCl + ZnI_2 = 2AgI + ZnCl_2$$

Care must be taken that the iodide is not in excess, or the following reaction will occur, cuprous iodide being precipitated—

$$2ZnI_2 + 2CuCl_2 = 2ZnCl_2 + Cu_2I_2 + I_2$$

and iodine liberated. The iodide is well stirred in, and the precipitate allowed to settle.

After the withdrawal of the liquor, the mud is moistened with hydrochloric acid and treated with zinc, when the nascent hydrogen reduces the silver iodide, and zinc iodide and metallic silver result.

$$Zn + 2HCl = ZnCl_2 + H_2$$
$$H_2 + 2AgI = Ag_2 + 2HI$$
$$ZnCl_2 + 2HI = 2HCl + ZnI_2$$

During the reduction the mass is kept warm by jets of steam.

The mud, or precipitate, after reduction, contains 6 to 12 per cent. of silver, a little gold, and a large percentage of lead and oxide of zinc, with sulphuric acid, lime, etc. The lead is reduced by the action of the zinc.

Von Patera's Process.—The solution of the chloride produced in chlorinating roasting by thiosulphate of soda, "hyposulphite," and precipitation of the silver as sulphide by sodium or calcium sulphides, was first proposed by Von Patera. Of late years, it has come prominently to the fore in a more or less modified form, and is the most important "wet" process for the treatment of *silver ores.*

In the American silver mills, where this process is pursued, the dried and crushed ore is chlorinated by roasting with salt.

After roasting—especially in White-Howell calciners—the ore is left for some hours in heaps, the chlorination proceeding after withdrawal from the furnace. It is then transferred to lixiviating vats, and leached with hot *water* to remove all soluble matters—zinc, manganese, copper, lead, and other chlorides—till the effluent liquor gives no precipitate with sodium sulphide. Some silver chloride is also dissolved. The stronger liquors from the first leachings are run into tanks, and the silver they contain is precipitated by the careful addition of sodium sulphide. It is thrown down before the other metals present are completely precipitated. This precipitate contains about 4 to 6 per cent. of silver.

The ore is then leached with sodium thiosulphate solution, of strength varying from $\frac{1}{4}$ to 1 per cent., according to the richness in silver. The solution is run by gutters under or alongside the tanks, into deep precipitation tanks holding about 1000 gallons (5 feet diameter and 8 feet deep), sodium sulphide solution is added, and silver sulphide precipitated thus—

$$(Ag_2S_2O_3,Na_2S_2O_3) + Na_2S = Ag_2S + 2Na_2S_2O_3$$

The regenerated thiosulphate solution is available for use again.

Treatment of Sulphide Precipitates.—The sulphide precipitates are roasted in a furnace, and, if poor in silver, smelted with lead, which decomposes the sulphide and takes up the metal.

The lead is cupelled to extract the silver. If the sulphide is pure, after roasting, it is melted in crucibles with carbon.

In roasting, and in treatment by lead, there is great liability to loss by volatilization and dusting. The flue dust from these furnaces assays up to 1200 ounces of silver per ton.

Formerly the silver precipitate was run down in crucibles with scrap iron, silver being liberated and iron sulphide formed. The regulus retained silver, and was re-treated. Calcium thiosulphate and calcium sulphide replace the soda salts in the "Kiss" process.

Treatment of Base Ores.—Ores containing much lead and zinc sulphides, antimony, arsenic, and bismuth, are unsuitable for treatment by the ordinary "hypo" process, the chlorinating and leaching being rendered difficult and incomplete in the presence of those bodies. Hence some of the silver remains in the mass as sulphide, and is not removed by hypo.

This difficulty is overcome in the **Russell** process, by following the ordinary thiosulphate leaching with a supplementary one by the double thiosulphate of soda and copper, formed by running the thiosulphate solution through a perforated box containing copper sulphate immersed in the leaching vat just above the ore. This method is rendered necessary by the decomposition of the double salt on exposure, and to prevent this the tanks are closed in. The double salt has the composition—

$$2Na_2S_2O_3, 3Cu_2S_2O_3$$

and the reaction is—

$$2Na_2S_2O_3, 3Cu_2S_2O_3 + 3Ag_2S = 2Na_2S_2O_3, 3Ag_2S_2O_3 + 3Cu_2S$$

The action of the *extra* solution is not rapid, and circulating pumps are employed to keep it in motion throughout the mass. Undecomposed sulphide of silver is thus dissolved out, and the silver in the residues is greatly reduced. The silver is precipitated by sodium sulphide as before.

The sulphide precipitates are, however, very impure, containing only 25 to 40 per cent. of silver. The excess of copper used in the *extra* solution is precipitated with the silver. Greater expense is entailed in refining in consequence. To obviate this, it has been proposed to treat the precipitate obtained from the *extra* solution with sodium nitrate and sulphuric acid, whereby the mixed sulphides are converted into soluble sulphates. The nitric acid fumes evolved are condensed, and the sulphur which separates used for making sodium sulphide.

The silver from the sulphate in the solution is then precipitated by copper, and the copper subsequently by iron.

In dealing with ores containing much galena, the lead sulphate and chloride formed in roasting dissolve in the hyposulphite. The lead is removed by the addition of sodium carbonate before precipitating the silver.

In zinc ores treated by this process, the zinc sulphate formed is dissolved out in the preliminary leaching with water.

In these processes, any gold contained in the ore is extracted to a large extent, and is precipitated with the silver as sulphide. During the roasting it, too, is chlorinated, and thus dissolved out.

The wooden tanks employed in these lixiviation processes are either round or square, well coated on the inside with tar. The capacity varies from 5 to 60 tons of material. They are provided with a perforated false bottom covered with canvas, on which a layer of filtering material, about a foot thick, is laid. This material consists of gravel and silver sand, in layers, or of sawdust, according to circumstances. Over the top of the filter is another canvas covering.

The leaching liquor is frequently poured on the top of the ore, but sometimes is introduced by a pipe below the false bottom, and allowed to percolate upwards until the mass is soaked, after which it is poured on top as usual. Below the false bottom is an opening in the side of the tank, by which the liquors are run off and carried by gutters into the settling and precipitation tanks. These, for convenience, should, if possible, be placed at a lower level. Steam-jet injectors are employed to elevate the liquors, if necessary.

Silver from Lead.—Pattinson's process for the concentration of the small amount of silver occurring in lead has been noted on p. 191, and the melting with lead of the roasted sulphide precipitates obtained in the Von Patera, on p. 223. Silver ores, if pure sulphides, are sometimes added to a bath of

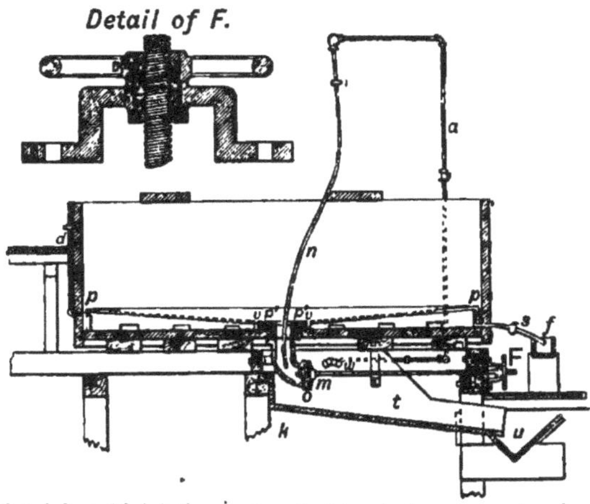

FIG. 79.—Self-sluicing Lixiviation-tank. *p*, false bottom; *a*, air pipe; *s*, pipe for removal of liquor; *f*, trough leading to precipitating tanks; *m*, plug for removal of residues; *n*, sluicing pipe for ejecting residues.

lead in a reverberatory furnace, much in the same manner as the poor Von Patera precipitates. The silver compounds are decomposed by the lead, and the silver passes into and alloys with the excess. A highly argentiferous lead is also obtained by the treatment of the zinc crusts removed in Parke's process for desilverizing lead (p. 197).

Cupellation of Rich Lead.—The lead is separated from the silver and gold by exposing the surface of the molten metal at a red heat to the action of a blast of air. The lead combines with oxygen, forming litharge (PbO), which melts, and is blown off the top, thus exposing fresh surfaces to the action of the air. Copper and other base metals present are also oxidized, and the oxides dissolved in the melted lead oxide and carried away by it. The silver and gold, which are unoxidizable, are left behind. Some little is, however, carried away in the oxide, particularly when the alloy becomes very rich. Bismuth remains until the last. In the English cupellation furnace, this oxidation is conducted on a bone-ash cupel, and some of the litharge is absorbed by the porous material. The bed of the German cupellation furnace is made of marl brasque—a mixture of marl, or clay and lime with wood ashes.

The English cupel or test is made by ramming bone-ash, moistened with a solution of pearl-ash, into an iron frame, with mallets. The frame, A, is elliptical in shape, 4 to 5 feet long and 2 feet 6 inches to 3 feet wide, made of 5-inch flat iron, from $\frac{1}{2}$ to $\frac{3}{4}$ inch thick. Five iron ribs, 3 to 4 inches wide and $\frac{1}{2}$ an inch thick, cross the bottom (Fig. 80). The bone-ash is rammed in in layers, and a cavity, E, scooped out with

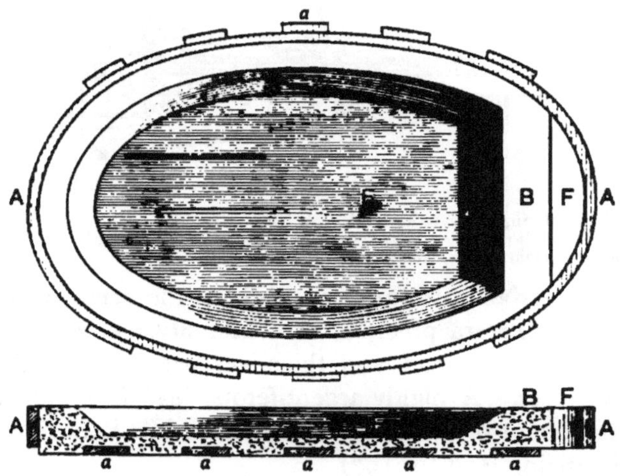

FIG. 80.

a trowel, leaving the sides about 2 inches thick round the top and 3 at the bottom, and the bottom itself about $1\frac{1}{2}$ inch thick. At one end, some 5 inches of bone-ash are left, and an opening, F, is made clean through the bottom, leaving a 2-inch dam, B. The litharge is thus prevented from coming into contact with the ironwork and corroding it. The cupel holds about 5 cwts. of lead.

This cupel forms the hearth of the cupellation furnace (Fig. 81). G is the fireplace, C the hearth, and B the stack. A tuyere, N, having a downward direction, enters at the back, and over the door is a hood, H, to carry away the fumes of PbO. P is a pot in which the lead is melted. Coal is used as fuel.

The cupel, carefully dried for some days, is placed on a truck, run under the furnace, and lifted into its place, in which it fits loosely. It is secured by wedges, crossbars, or by projecting eyes, and the edge of the iron ring covered with bone-ash. After carefully heating to redness, lead is introduced from the lead-pot, or in pigs, through a channel at the back. The blast is supplied by a fan, or occasionally by a steam jet. The litharge which forms is removed by making a little gutter in the bridge in front, through which it flows into conical iron moulds on wheels, placed beneath to receive it. The temperature of the furnace is cherry redness. As the lead is removed by oxidation, fresh additions are made to keep up the level in the cupel.

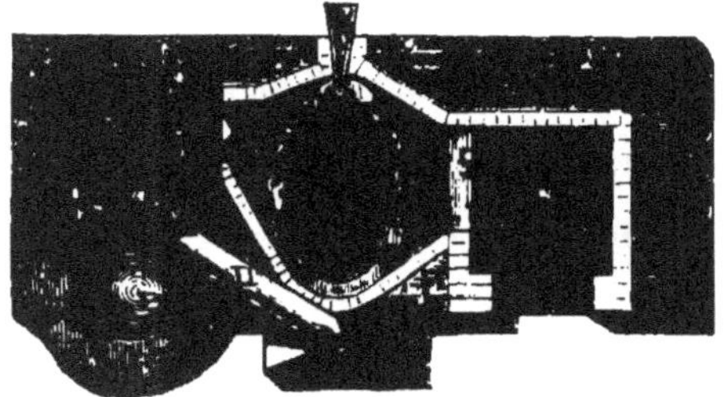

FIG. 81.

In working on Pattinson lead containing from 500 to 700 ozs. to the ton, the operation is conducted in two stages. In the initial stage a lead containing 8 per cent.—4000 to 5000 ozs. per ton—is produced. The litharge produced in this stage is poor enough in silver to be sent into the market. It is sold for glass-making, etc. The enriched lead is then generally removed, being run into pigs by boring a hole through the bottom of the cupel. More poor lead is then treated in the same cupel after stopping the hole.

The rich lead is then similarly treated on a new test, the

litharge being saved separately. It is reduced as described, (page 191), and yields lead containing some 40 ozs. of silver per ton. As the cupellation approaches completion the surface of the metal becomes iridescent (rainbow tinted) and strikingly beautiful. As the last film of oxide clears off, the metal flashes out brightly, presenting a clear, brilliant, bluish-white appearance, the surface reflecting the roof of the furnace. This is known as the "brightening" or "coruscation." The cooling of the silver must be effected slowly to prevent loss by "spitting." This, as already indicated, is prevented by a small amount of impurity, and its occurrence is an index of the purity of the metal. Many curious and fantastic forms result by the throwing up of the surface, partly caused by the escape of oxygen and partly by contraction of the mass expelling the fluid interior.

In an ordinary furnace from 4 to 5 cwts. of lead are oxidized per hour, some $1\frac{1}{2}$ cwt. of coal being required.

The silver is generally about 995 to 998 fine. The cupels are broken up and the parts saturated with litharge smelted with fluor spar in a blast furnace to recover the lead.

Electrical Refining.—In Keith's process, the rich lead is made the anode (dissolving pole), and a sheet of pure lead the cathode. A solution of lead sulphate in acetate of soda is employed as the electrolyte. The depositing tanks are arranged in series, and a strong current is employed. The anodes are enclosed in muslin bags, and, as they dissolve, the precious metals and other insoluble matters are retained. The lead is deposited in a crystalline or pulverulent form, and falls to the bottom of the tanks, from which it is removed, compressed and melted. It carries about half a pennyweight of silver to the ton. The residues in the bags are cupelled with lead.

Refining.—The refining of impure silver is effected either by cupellation, or, if very impure, such as is sometimes obtained by amalgamation methods, by melting it and exposing it to the air in crucibles. Copper, iron, etc., may thus be largely removed as dross. The purified metal is then refined on bone-ash cupels.

Separation of Silver from Copper.—Formerly a method of separating silver from argentiferous copper by means of lead was largely practised. The copper, melted with about four times its weight of lead, was cast into flat cakes 18 inches in diameter and 3 inches thick. These were then carefully heated and the lead allowed to liquate out, carrying the silver with it. The residues were subjected to a second liquation at a higher temperature. The Argentiferous lead was afterwards cupelled.

CHAPTER XVI.

GOLD.

THE fine yellow colour and brilliant appearance of this metal are well known. It is comparatively soft, being only slightly harder than lead when pure and unalloyed with base metals. It is the most malleable and ductile among metals, leaf $\frac{1}{280000}$ of an inch thick being obtained by hammering, and a grain can be drawn into wire 500 feet long. Its tenacity is about 7 tons per square inch. These properties are influenced greatly by minute quantities of impurity, notably lead, bismuth, antimony, and arsenic. Its alloys with silver and pure copper are harder, but extremely malleable and ductile. It has a melting-point of about 1075° C., and volatilizes at very high temperatures, as in an electric furnace. When molten it appears greenish, and if undisturbed as it cools, suddenly flashes out bright when at a temperature of about 600°, after which it cools and solidifies. It contracts considerably on solidifying. Pure gold welds with the greatest ease. Its flowing power is high, and it is an excellent conductor of heat and electricity. Its specific gravity is 19·3.

The metal is unaffected by dry or moist air, and resists the action of acids (save selenic), alkalies, and sulphuretted hydrogen. It is readily attacked by chlorine, and also by iodine and bromine. A mixture of nitric and hydrochloric acids (aqua regia) dissolves gold, because free chlorine is generated. The gas attacks gold most rapidly at the moment of liberation (nascent state), and is less active when diluted with air or any inert gas. The chloride ($AuCl_3$) is very soluble in water. It is decomposed at high temperatures, gold and chlorine resulting.

Gold is slowly dissolved by cyanide of potassium in presence of air or oxygen.

$$2Au + 4KCN + H_2O + O = 2AuCN.KCN + 2KHO$$

The addition of a little bromine or cyanogen bromide hastens the solution. It is precipitated from its solution as chloride by most metals, and also by sulphate of iron, chloride of antimony, oxalic acid, carbon, and carbonaceous bodies. The solution of gold in potassium cyanide is not precipitated by ferrous sulphate or ordinary reducing agents. Metals, as, for example, zinc, throw it down readily. Mercury readily amalgamates gold.

Occurrence.—Gold occurs naturally in the free state, but to some extent also as telluride, and possibly as sulphide. It is intimately associated with iron pyrites and other sulphides, the greater part of the gold in some ores being contained in the pyritical contents. Native gold occurs in matrix, generally in veins of quartzose and other hard rocks, forming reefs or dykes, and in deposits formed of the débris produced by the weathering of such rocks, such as "alluvium" river sand, etc. In the latter formations, owing to the action of running water, the lightest portions have been transported furthest, and an accumulation of the coarser gold has taken place nearest to the broken-down rock, owing to the high specific gravity of the metal. Alluvial deposits are, in consequence, often richer than the mother rock from which they are derived. The gold occurs in all degrees of coarseness, from microscopic particles to masses of considerable size. These are known as "nuggets." The "Maitland Bar" nugget from New South Wales, exhibited at the Mining Exhibition of 1890, contained 313·093 ozs. of fine gold. Gold is very widely distributed in small quantities.

In Great Britain, it has been found in Cornwall, Wales, Perthshire, and Sutherlandshire; in Ireland, in Wicklow, and in the Isle of Man.

In Europe, Hungary, Transylvania, Sweden, Spain, and Italy also furnish gold.

Rich deposits occur in India, Ceylon, China, Japan, Siberia, the Ural Mountains, and in South Africa.

In the New World, the gold-bearing rocks occupy the west coast, following the line of the mountains. Alaska, British Columbia, California, Mexico, Bolivia, Peru, Chili, Columbia, and Brazil are all rich in gold. Australia is at present one of the principal gold-producing countries.

The high value of the metal enables deposits which contain very little gold—in some cases only a few grains per ton—to be profitably worked. Much depends on the nature of the deposit and the method adopted.

Alluvial Deposits, Placers, etc.—The mining and extraction of gold are almost inseparable. Alluvial deposits differ very much in character, from loose sand, pebbles, etc., through stiff earth, to hard conglomerate, the pebbles of which are firmly cemented together. The "banket ore" of South Africa is of this class, although the pieces are angular. It appears to be more of a "breccia."

The gold in alluvium occurs in nuggets of varying size, and in grains. In surface deposits (placers), generally shallow, the ground is first picked over for nuggets. The sand and gravel are then washed, the lighter materials being thus removed, and gold remains.

Panning out consists of washing the "pay-dirt" in a shallow iron pan with a depression in the middle for retaining the gold. The earth is placed in the pan and washed under water, a circular motion being given to it. Light matters are carried over the edge, and the gold gradually finds its way to the bottom, together with other heavy matters. This residue is dried, and the lighter materials blown away, leaving the gold. In Africa the natives wash the river sand in gourds, mixing it up with water and pouring off the matter held in suspension, and store the gold dust obtained in quills.

Hydraulic Mining.—In this method of working, the auriferous gravel is dislodged by means of a powerful jet of water directed against the bank by means of an iron nozzle (monitor). The quantity of water required for this purpose is enormous. It is sometimes carried for miles down hillsides and across valleys, in pipes on trestles (flumes), and is delivered at high pressures, sometimes under a head of 200 feet. The dislodged material is carried by the stream of water down a series of long wooden troughs, the "sluice," made in 12-feet lengths, fitted together, which have a slope of about an inch to the foot, more or less. The bottoms of these are crossed at intervals by movable wooden or iron bars, *riffles*, behind which the heavy particles of gold, which move more slowly and have a greater tendency to settle, lodge. The lighter gravel, etc., is carried on by the current. Perforated iron plates, *grizzlys*, are introduced at intervals in the bottom of the sluices. The coarse gravel is carried on over these, but the finer portions fall through the plate on to a second sluice at a lower level, with a separate water supply. The inclination of this is less than the first, and the lower velocity of the stream favours the collection of the fine particles.

Mercury is fed in small quantities from time to time at the top of the sluice. It lodges behind the riffles and catches the particles of gold which come in contact with it. Amalgamated copper plates are often suspended in the sluice to catch the "float gold" (very small flattened particles which float on the surface).

The amalgam is removed at intervals. For this purpose the water is stopped, the gravel cleaned out and the riffle bars lifted in order, commencing at the top, and the amalgam collected. The upper part of the sluice is cleaned up at frequent intervals, the greater part of the gold being caught there. After squeezing out the excess of mercury through chamois leather, the amalgam is retorted.

Washing Sands, etc.—Baize, blankets, and hides, with the hairy side up, are sometimes employed in the washing of fine sands and stamped material. They are attached to sloping boards which form the bottom of shallow sluices. The sand is thrown on at the top, washed down by a stream of water, and brushed by a workman against the stream. From time to time the blankets, etc., are removed, and the gold, etc., shaken out into a trough of water and amalgamated with mercury. The amalgam is afterwards retorted.

A very effective method of treating fine sands is to boil

them with water and mercury. Chinamen find it profitable to work over the "panned-out placers" in California in this way.

Hard "cements" (consolidated alluvium) are often ground in mills resembling mortar-mills, the "pulp" or ground material being carried away over amalgamated copper plates.

Treatment of Gold Quartz.—Much depends on the mode of occurrence of the gold and the nature of the quartz. In some ores it occurs entirely in the free state, and free from pyrites. These are often ferruginous, the oxide of iron resulting from the decomposition of iron pyrites. The gold often exists in largest quantity in the oxide of iron, showing it to have been derived from the pyrites. Such ores generally become pyritous below the water line. "Gossans" consist mainly of this decomposed pyritous material.

In quartz containing pyrites, much of the gold is often contained in the pyrites. Most of this escapes extraction by the ordinary amalgamation processes, either owing to its extreme state of division, or to combination as sulphide in the pyrites. It passes into the "tailings," as the residues from amalgamation are called. Such ores are described as "refractory," and require special treatment. Ores in which all the gold is recovered by simple crushing and amalgamation are described as "free milling."

Amalgamation of Free-milling Ores.—The quartz is first broken into about inch cubes in a stone breaker or ore crusher (Fig. 82), the jaws being so adjusted as to deliver it of the required size.

The ore is next crushed to fine powder under stamps, or by rolls, or grinding mills.

Fig. 83 shows a stamp battery. The stamps consist of heavy cast-iron "heads," or "bosses," A, shod with steel, attached to long wrought-iron or steel stems sliding in hardwood "guides," BB. These are lifted by the "cams," C, attached to the revolving "cam shaft," D, driven by the pulley, E, acting on the "tappets," F, keyed on the stems. The cams are right- and left-handed, so that each head is raised twice at each revolution. Under each stamp is a steel-faced "die," G, between which and the falling head the ore is crushed. The dies are contained in the cast-iron "mortar-box," H, which is supported on a wooden foundation on pads of indiarubber a quarter of an inch thick. One or both sides of the mortar is fitted with "screens," I, of perforated sheet iron or thick wire cloth, and a stream of water is fed in from a pipe. Some 72 gallons per hour per head is required, but it may be recovered in settlers with a loss of about 25 per cent. The

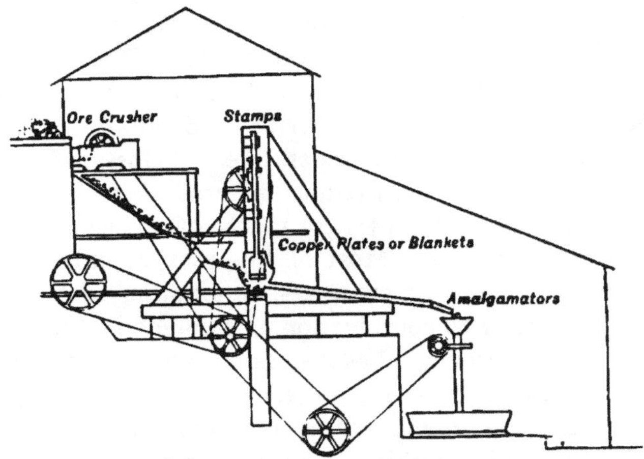

FIG. 82.—Wet-crushing Gold-mill.

ore is fed in on the side opposite the screens (if single discharge), often by an automatic contrivance. The action of the cams on the tappets not only raises the head, but turns it partially round each stroke, and thus causes it and the die to wear uniformly. For gold crushing, the mortar-box is lined with amalgamated copper plates, and much of the coarse gold is caught in the mortar. A little mercury—half a thimbleful—is fed in at intervals for this purpose. Levers on "jack" shafts, K, are provided for holding up the stamps.

The heads, with attachments, weigh from 4 to 9 cwts., but for quartz, generally about 7 cwts.

They have a drop of about 10 to 12 inches, and give from 70 to 80 blows per minute, the cam shaft making from 35 to 40 revolutions per minute. The fall of the head splashes the pulp against the screen and assists in forcing it through.

The screens have a mesh of from 30 to 60 per linear inch.

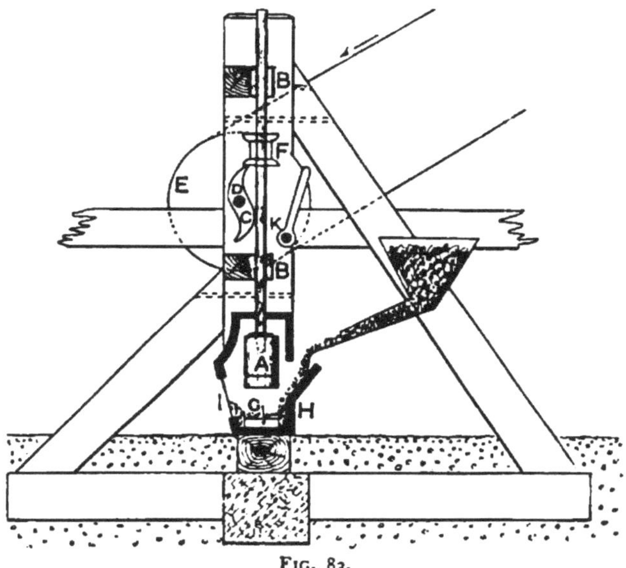

FIG. 83.

Each head will crush 2 to $2\frac{1}{2}$ tons per day (wet crushing).

FIG. 84.—Stamp Battery.

The fine material, "pulp," is carried by the water through

the screens, on to amalgamated copper plates (Fig. 84), slightly inclined, over which it is carried by the current, the free gold dragging along the bottom and being arrested by the mercury.

The plate next the mortar is sometimes silver-plated, to prevent the deadening of the mercury by oxidation of the copper dissolved in it, making it much less active.

From the plates the "tailings" may pass to amalgamators or to vanners, as subsequently described.

Stamp batteries, "dry" or "wet," are open to several objections. The principal one is its tendency to cut up and flatten out coarse particles of gold. By the repeated hammering a hard surface is developed, and fine particles of foreign matters driven into the soft metal. This renders it very difficult to amalgamate, the mercury only attacking it with extreme slowness. There is also liability to loss from the production of "float gold."

FIG. 85.

Rolls are open to fewer objections, cracking open the ore and exposing the metal. Fig. 85 shows the Huntingdon Mill. The pan is of cast iron, with a steel belt round the lower part inside. Four mullers are supported vertically on rods, on which they revolve by friction against the side of the pan. The head from which they are carried is revolved at 70 revolutions per minute. The mullers are pressed by centrifugal force against the ring, and the ore coming between is crushed. Above the steel roller path is a screen occupying half the circumference of the pan. A stream of water is fed in above, and stirrers are provided to ensure complete crushing. For soft ores a mill having a 5-foot pan is about equal to a 10-head stamp battery, and requires only about half the power to drive it.

Cleaning up.—The mill is stopped periodically for the purpose of collecting the amalgam which is carefully removed from the amalgamated plates. It is worked up with fresh mercury either by hand or in pans (clean-up pans), and well washed with water to remove earthy and other matters. It is then squeezed in bags of canvas or chamois leather to expel the excess of mercury. This is not free from gold, but is re-used. The pasty amalgam remaining in the bag is retorted to remove the mercury. The gold which remains is melted in crucibles and refined.

Loss of Mercury.—This arises from two causes, "sickening" and "flouring." In the former case, the mercury is converted into a black powder and carried away. It is caused by the presence in the ore of certain minerals, *e.g.* antimony sulphide.

"Flouring" is breaking up of the mercury into minute globules, which are lost.

Treatment of Tailings in Amalgamators.—The slimes, or tailings, may be passed into amalgamators, in which they are ground up with mercury (Fig. 86).

The pulp is fed into the hopper, A, and passes down the hollow shaft, B, to which is attached the iron muller, C. This is slowly rotated; the outer pan, D, contains mercury, below which the muller dips.

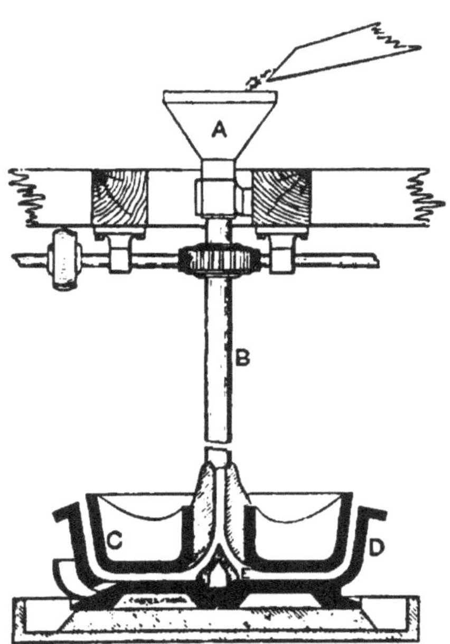

Fig. 86.—Continuous Amalgamating Pan.

The pulp is delivered, by the openings E, *under the muller*, which, by its revolution, thoroughly incorporates the tailings with the mercury. They escape over the edge of the pan, and may be delivered to a second set of amalgamators at a lower level, or pass direct to settlers.

The Hungarian mill for the amalgamation of iron pyrites is in principle exactly similar, but is driven from below. Amalgamators of other types are also employed, *e.g.* Berdan pan.

The tailings are more commonly treated on Frue vanners or other contrivances for the recovery of the heavy sulphides—

iron pyrites, galena, copper pyrites, etc.—which they contain, and which often carry a considerable portion of the gold present in the ore. This is not recovered by simple crushing. The "concentrates," as they are termed, are either ground in iron pans with mercury, as in the treatment of silver ores (p. 214), or are roasted and treated by chlorination (see below). In some cases the whole of the pulp is treated *without concentration* with potassium cyanide (see p. 241).

Chlorination Processes.—As noted (p. 229) gold is readily attacked by chlorine gas, and the chloride formed is soluble in water. In 1853 Plattner proposed to extract gold in this way. Only ores in which the gold is free can be thus treated. It is employed for the treatment of the pyritical concentrates obtained from the tailings.

They are first roasted to remove sulphur and open up the ore. This converts the iron into ferric oxide—a form in which it is not acted on by the chlorine—and leaves the gold free.

This is effected in a Brückner, or other calciner. The roasted ore is moistened and put into well-tarred vats provided with false bottoms and tightly fitting covers, with a bung-hole in the top. When full, the vats are covered and luted, the bung-hole being left open. Chlorine from a generator (see Fig. 86) is admitted under the ore, and, when it has displaced the air, and is escaping freely, the opening in the top is closed, and the gas allowed to operate for some 24 to 72 hours. The excess of chlorine is blown out, and the chloride formed dissolved out with water. The solution, after settling, is run into the precipitating tanks, which are lead lined, and provided with stirrers, and sufficient ferrous sulphate is added to precipitate the gold.

$$6FeSO_4 + 2AuCl_3 = 2Fe_2(SO_4)_3 + Fe_2Cl_6 + Au_2$$

The precipitate is allowed to settle and the clear liquor siphoned off through a sawdust filter. The solutions obtained from several batches of ore are usually treated before gathering the precipitated gold. After treatment with acid, to dissolve out any basic salts of iron which may have been formed, it is fused up. Its purity varies from 920 to 990 parts of gold per 1000.

The chlorine is generated in a lead still from salt, manganese, dioxide, and sulphuric acid.

Modern chlorinating vats are generally mounted on trunnions to facilitate the removal of the residues.

Many modifications of the process are in use. The improvements principally relate to—

1. Chlorination of the material under pressure:
   - (*a*) Of the gas itself.
   - (*b*) Of air pressure (Newberry Vautin).
   - (*c*) Hydraulic pressure (Pollok).

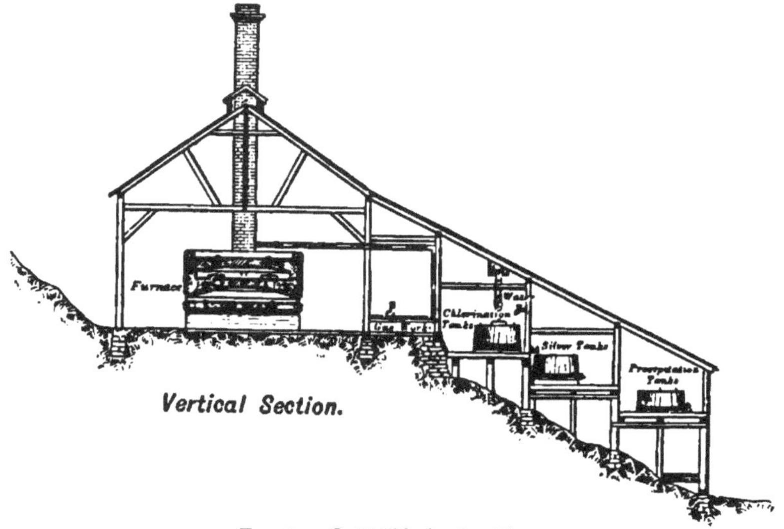

FIG. 87.—Gold Chlorinating Plant.

2. Agitation of the ore during chlorination:
   - (*a*) By agitators (Cobley and Wright, De Lucy and others).
   - (*b*) By revolving barrels (Duflos, Primard, Mears, Newberry, Vautin, Pollok, etc.).
3. Filtration by suction and centrifugal force (Aarons, Newberry and Vautin), etc.
4. Generation of chlorine in the vats by the action of sulphuric acid on bleaching powder mixed with the ore, and absorption of the excess of chlorine by passing it through milk of lime.

5. Heating up liquors to expel excess of chlorine prior to precipitation.

6. Precipitation by charcoal, bitumens, and other reagents (the charcoal filters are burnt and the ashes fused with borax to recover the gold).

By these means the time occupied is shortened.

Extraction of Silver.—When the ore or concentrates contain silver, a little salt, about 1 per cent., is added towards the end of the roasting, and the chloride of silver formed is dissolved out by hyposulphite of soda after the removal of the gold chloride (see p. 222).

The Cyanide Process, introduced by Messrs. MacArthur and Forrest, has already been extensively adopted, and promises to displace chlorination processes, if a cheaper method of making cyanide can be found.

It depends on the well-known fact that potassium cyanide attacks gold in the presence of oxygen, and if in a fine state of division, rapidly dissolves it. Weak solutions are found to be more active than strong ones, on account of the greater solubility of oxygen in them (*Journal of Chemical Society*, 1893).

One great advantage possessed by this process is that raw pulp may be treated directly, no previous calcining or concentration by washing on vanners or buddles, etc., being necessary.

The strength of solution varies from 0·4 to 1 per cent., but 0·5 is said to be efficacious. The tailings, or concentrates, are left in contact with the solution for from 60 to 72 hours, circulating pumps being employed. The clear liquid is then run through boxes containing zinc shavings. The gold is precipitated by the zinc as a black powder. It is collected periodically, washed to free it from zinc, as far as possible, and melted in crucibles. A very coarse bullion is obtained, and slag. This is melted with lead and cupelled, to recover the gold it contains. About 90 per cent. of the gold is obtained, and the liquors may be re-used. The foreign matters in the ores are not attacked.

Refractory ores which cannot be successfully treated in any other direct way may be thus dealt with. The process is largely used in South Africa. It is not applicable to coarse gold, and is used for tailings and concentrates only. The danger arising from the use of such a powerful poison as cyanide, is very small, if cleanliness and perfect ventilation are attended to.

In the Siemens-Halske process the gold is precipitated by electro-deposition.

Sulman adds cyanogen bromide to the ordinary solution, which greatly reduces the time, and is more effective. The precipitation is effected by zinc fume instead of shavings (see p. 261).

Tailings containing much pyrites (especially copper) are difficult to treat. On exposure to the atmosphere in the moist state, they oxidize. Free acid and soluble salts are formed, which greatly increase the consumption of cyanide. Lime is used to neutralize the ore and decompose soluble sulphates.

Parting.—Native gold and bullion almost invariably contain silver and other metals from which it must be separated. The base metals may be removed by cupellation, or, if fairly pure, by fusion with nitre and borax; but silver, platinum, etc., will remain, and must be removed by chemical means. This process is known as "parting," and consists of dissolving out the silver by acids.

Alloys of even base metals with gold, are not attacked unless the base metal is in large excess; and hence, if a sufficient proportion of silver is not present, enough must be added to ensure it being acted on.

Sulphuric Acid Parting.—Silver is soluble in hot, strong sulphuric acid, forming sulphate of silver. The alloy to be parted must contain not less than 80 per cent. of silver; generally the alloys treated contain very much more. The metal is granulated by pouring it into cold water, so as to offer a large surface to the action of the acid.

The parting pans are usually of white cast iron, about 2 feet wide, provided with a lid and a pipe by which the SO_2, generated, is conveyed into a lead chamber for reconversion into sulphuric acid, to be re-used.

$$Ag_2 + 2H_2SO_4 = Ag_2SO_4 + 2H_2O + SO_2$$

The pots are heated by fires from below. The granulated

metal is treated with about $2\frac{1}{2}$ times its weight of strong acid, at a temperature about the boiling-point of sulphuric acid. The sulphate of silver which forms separates in a pasty mass of small crystals. This is thrown into water in a lead-lined tank and heated by steam. The sulphate dissolves, and the gold settles out and is washed and collected. It still retains silver, and is next treated with sulphuric acid and sulphate of soda in the proportions of 3 to 5, and strongly heated.

The sulphate raises the boiling-point of the acid, and enables it to attack the remaining silver. A second treatment is sometimes necessary. What remains in the pans is then boiled with acid, and the residue washed, dried, and fused.

The solution of sulphate of silver is decomposed by copper ($Ag_2SO_4 + Cu = CuSO_4 = Ag_2$), and the precipitated silver compressed by hydraulic pressure, dried, melted in crucibles, and cast into ingots. The copper is afterwards deposited by iron; or the solution is concentrated, and the copper sulphate allowed to crystallize out, and sent into the market, the mother liquor being further concentrated in glass or platinum vessels, to recover the excess acid for re-use.

Silver containing more than three grains of gold per pound can be economically treated by this method. Much old silver plate was sacrificed for the gold contained in it when this process was introduced, the older method of parting by nitric acid being too expensive.

Parting with Nitric Acid.—In this process, nitric acid is substituted for sulphuric acid. The operation is conducted in platinum, glass, or porcelain stills, with covers connected with condensers, to recover the acid which is boiled off. Nitric acid does not readily attack the alloy unless it contains about three times as much silver as gold. If less is present, it is melted with more silver, to make up that amount. This is known as *inquartation*. The alloy is granulated and boiled with about twice its weight of nitric acid, diluted with one-third its bulk of water. Red fumes are evolved as long as silver is dissolving.

$$6Ag + 8HNO_3 = 6AgNO_3 + 4H_2O + 2NO$$

The solution of nitrate of silver is drawn off, and the

residual gold treated with a little strong nitric acid, and after washing, melted under borax and cast into ingots.

The silver is recovered by adding hydrochloric acid to the solution of silver nitrate, which precipitates the silver as chloride.

The acid is added cautiously so as to leave a little silver unprecipitated in order that the nitric acid formed may be re-used.

$$AgNO_3 + HCl = AgCl + HNO_3$$

If free HCl were present in the acid used for parting, the gold also would be attacked by the chlorine generated. It is detected by the addition of silver nitrate to the acid.

The silver chloride is reduced by zinc.

Separation from Platinum.—In parting with nitric acid, platinum, if present to a less extent than 9 per cent., is dissolved out with the silver. If much platinum is alloyed with the gold, it is best separated by solution in aqua regia and precipitation of the platinum by sal-ammoniac.

Alluvial gold often contains grains of a heavy, hard alloy of osm. iridium, which is not taken up by the gold. In the American Mint, this is separated by melting the metal in tall crucibles, when it sinks owing to its greater specific gravity. Silver is also alloyed with the metal to lower its specific gravity and permit the osm.-iridium to sink more rapidly. The upper layers are ladled out and parted, and a fresh batch treated. The residue at the bottom is remelted with silver several times to diminish the gold present, and then parted with nitric acid. The silver dissolves, and the grains of osm.-iridium remain mixed with a little pulverulent gold, which is separated by washing.

Toughening Brittle Gold.—Mere traces of arsenic, antimony, bismuth, and lead suffice to render gold brittle. It is toughened by treating the molten metal with mercuric chloride, or by passing chlorine through it by means of a clay tube, as practised at the Royal Mint (Miller's Process). The bismuth, arsenic, and antimony chlorides volatilize. If silver is present the silver chloride formed fuses and rises to the top. The gold is not attacked, gold chloride being decomposed at high temperatures.

Smelting with Lead.—Old crucibles that have been used for gold and silver melting are first picked over, ground, and amalgamated. The residues are then smelted with lead-yielding materials, and the resulting metal cupelled to extract the gold. "Sweep" (sweepings) is similarly treated. Ores are also sometimes smelted with lead.

Alloys.—The common method of expressing the quality of a gold alloy is in "carats" and carat grains (4 c.g. = 1 carat).

Pure gold is 24 carats fine; 18-carat gold contains $\frac{3}{4}$ gold and $\frac{1}{4}$ alloy, or 750 parts per thousand; 9-carat gold, 375 parts per thousand. English-coinage gold is 22 carat, or 916·6 parts per thousand. Its specific gravity is 17·157. The alloy is copper, which is added to harden it. A new sovereign weighs $123\frac{1}{4}$ grains, but is legal tender so long as it does not fall below $122\frac{1}{2}$ grains, $\frac{3}{4}$ grain being allowed for wear. It is estimated to circulate for 18 years without becoming light. The weight of metal in the coin is worth its face value. The French and United States standard alloy is 900 fine, or 21 carats $2\frac{3}{8}$ carat-grains. The terms "worse" and "better" are applied as in silver. Thus the above alloy is 0 carats $1\frac{5}{8}$ carat-grains worse than English gold.

Articles of gold jewellery down to 9 carats are stamped by the Goldsmith's Company with a Hall-mark, indicating the quality, the year of production, and the Assay Office at which the tests were made.

Copper and silver are usually alloyed for the purpose of hardening gold, when malleability is required for purposes of working. Zinc is sometimes added when rigidity and hardness are most important. Pencil-cases and watch-guards often contain the latter metal.

CHAPTER XVII.

TIN.

Physical Properties.—Tin is a white metal with a faintly yellowish tinge. It has a high lustre, and is very malleable. Foil, $\frac{1}{1000}$ of an inch thick, can be obtained by beating. It is ductile, but its tenacity is low—only about 2·1 tons per square inch. Its melting-point is 230° C., and it is not sensibly volatile at furnace temperatures if closely covered to exclude air. Near the melting-point it is brittle, and a cake of tin heated till the edges begin to melt and then dropped on the ground, breaks up into peculiar long-shaped, columnar pieces known as "grain tin." Tin which is impure does not readily do this. When bent, a strip of tin emits a peculiar crackling sound known as the *cry*. This is supposed to be due to the internal friction between the crystalline particles.

Tin is readily obtained in crystals, like antimony or bismuth. If the surface of an ingot is treated with a mixture of sulphuric and nitric acids, beautiful crystalline markings make their appearance. This is known as the *moirée metallique*, and is used as a metallic ornamentation, being coated with coloured varnishes. The metal is a poor conductor of heat and electricity.

Pure tin, when cast in a mould, at a low temperature, solidifies with a bright metallic appearance; but if impure, it presents a more or less frosted appearance, according to the amount of impurity present.

Commercial tin often contains small quantities of lead, copper, arsenic, antimony, and tungsten.

The metal is not affected by dry or moist air at ordinary temperatures. Heated in air, it oxidizes, forming stannic oxide (SnO_2). It combines readily with sulphur, forming SnS. This when roasted does not form sulphate, but yields SnO_2 and SO_2. It is decomposed when heated with metallic iron. Tin dissolves in hydrochloric and sulphuric acids. Nitric acid acts violently on it, and converts it into an oxide. It is also soluble in caustic soda and potash, forming stannates.

It is not readily attacked by vegetable acids or animal juices, and tin plate is hence largely employed in the canning industry. For the same reason it is used for coating the interior of vessels for cooking purposes.

Ores of Tin.

Cassiterite—Tinstone (SnO_2).—This is the only important ore of tin. It is yellowish-brown or black in colour, and occurs crystallized, in well-defined veins, and in grains, sometimes distributed through a mass of rock, such as granite. It has a specific gravity of 6·5 to 7. It has a high lustre, and is too hard to scratch with a knife. In the vein it is associated with galena, blende, copper, and iron pyrites, arsenical iron pyrites, and other minerals. Wolfram (tungstate of iron), another remarkably heavy mineral, is also associated with it. A vast number of non-metallic minerals occur as associates. Fluor, garnet, mica, chlorite, granite, gneiss, and porphyry may be mentioned. *Stream tin ore* is tinstone which has accumulated by the weathering of the rocks containing it. The lighter portions have been removed by the action of running water, and the tinstone and associated heavy minerals left. *Wood tin* is tinstone showing concentric markings more or less resembling wood. Tin ores are often very impure, sometimes not containing more than 1 per cent. of tinstone. Its high specific gravity facilitates dressing operations, and such ores, by careful picking, stamping, and washing, can be profitably worked. Tin ore is found in Cornwall and Devon, Germany, Spain, Russia, Malacca (Banca), Australia, United States, and Mexico.

The ore is carefully dressed by hand picking, stamping, and washing, to remove gangue. The copper and arsenical pyrites are not completely removed, and the wolfram also remains with the tinstone.

Bell-metal ore, or tin pyrites, is a mixture of copper, iron, and tin sulphides.

Smelting.—The ore is first carefully calcined in a large, low, reverberatory furnace, being turned over every 20 minutes or so. In Brunton's calciner the bed is circular, and revolves on a vertical axis, the turning over being done mechanically. In roasting tin ores a moderate heat is necessary at first, to avoid clotting of the sulphides present.

During roasting the arsenic combines with oxygen, and is converted into white arsenic (A_2O_3), which is volatilized and deposited in long flues provided for that purpose, from which it is collected. Sulphur burns off as SO_2, and the copper is largely converted into sulphate.

The ore after this roasting is moistened, and left in a heap for a few days, to allow more soluble sulphates to form. It is then thrown into a tank and agitated with water. Copper sulphate and other soluble matters are dissolved, and the sediment consists mainly of stannic and ferric oxides. The lower layers contain the larger proportion of oxide of tin owing to the greater rapidity with which it settles, on account of its high specific gravity. The ferric oxide is separated by washing, and the concentrated oxide is known as *black* tin. It is sorted into various grades according to purity.[1]

Reduction.—In this country this is effected by heating the tin oxide mixed with carbon in the form of anthracite, in reverberatory furnaces, of the form shown in Fig. 88. The bed measures about 15 feet by 9 feet, and slopes towards the tap-hole *o*, outside which is the "float," or tin-pot; which is lined with clay to prevent the metal taking up iron. The stack is 40 or 50 feet high. The bed is of fire-clay resting on slate slabs supported by iron bars, and the fire-bridge is about 14 inches high.

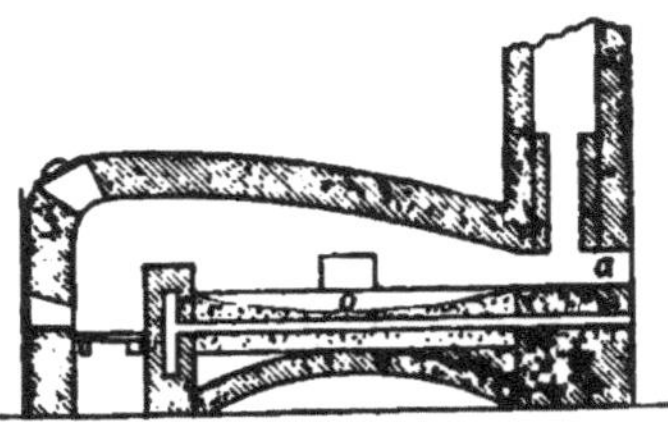

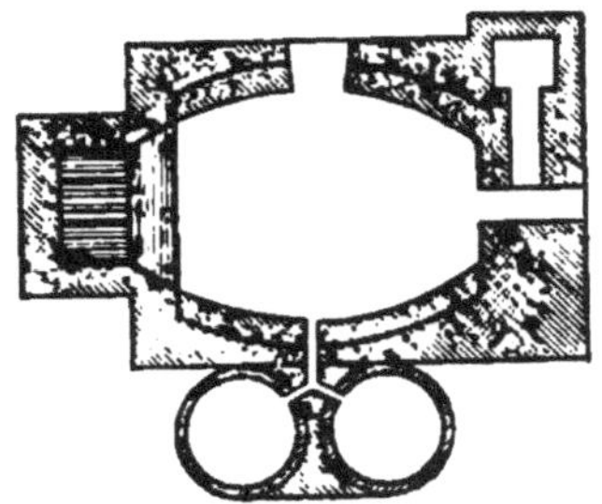

FIG. 88.—Furnace for smelting Tin Ores.

The charge consists of about a ton of black tin mixed with about 3 to 4 cwts. (20 per cent.) of anthracite powder, according to purity. If silica is present a little lime or fluor spar is added as a flux. The mixture is damped to prevent dusting, and after introduction into the furnace, the doors are closed and luted round. A low temperature is maintained for some time to

[1] **Removal of Tungsten.**—In the preliminary calcination wolfram is not affected. It is removed when necessary or desirable, by mixing the black tin with sufficient carbonate or sulphate of soda and heating in a special furnace to decompose the wolfram, yielding tungstate of soda and oxide of iron. The tungstate of soda is dissolved out by water and crystallized, and the oxide of iron removed by washing as before (Oxland's Process).

ensure reduction of the tin oxide and prevent the formation of silicate. In 4 to 5 hours the charge is well stirred up, anthracite culm thrown in, and the charge again heated for another hour. After again rabbling, the metal is allowed to subside, and is then tapped into the float.

The reduction takes place as follows—

$$2SnO_2 + 2C_2 = Sn_2 + 4CO$$

A fluid slag, known by the smelter as "glass," runs from the furnace with the metal. It consists of silicates of iron, lime, and alumina. Oxide of tungsten is often present. It sometimes contains as much as 20 per cent. of tin. It is allowed to accumulate, and resmelted to extract the metal.

A pasty mass, consisting of shots of tin, anthracite, and slag is left on the furnace bottom, and is raked out by the smelter. The tin contained in it is separated by stamping and washing.

The metal is ladled from the tin-pot into pig moulds, or, if pure, is run into the "boiling" pot.

Reduction in Blast Furnaces.—Tin ore is also reduced in small blast furnaces, with charcoal as fuel. The materials are charged in at the top, and the furnace works continuously.

The loss of tin in the slags is much greater, but the tin obtained is very pure.

Smelting in blast furnaces is abandoned in this country, but in Saxony, the East Indies, and other places is still continued. About 32 cwts. of charcoal per ton of tin are consumed in smelting.

Refining.—This involves two operations—liquation and boiling.

Liquation.—The pigs of tin, which weigh from 3 to 4 cwts. each, are piled on the hearth of a reverberatory furnace, with a bed somewhat more sloping than the reduction furnace, and heated carefully to the melting-point of tin. Some 18 tons of pigs are treated at once. The temperature is very carefully regulated. The purer tin melts, drains away, and flows out into the refining kettle. The impurities remain in an unfused state on the hearth, as a metallic, yellowish white, hard, brittle mass, often porous, known as *hard head.* It contains iron, tin, arsenic, sulphur, and a little copper. Sweating at a higher temperature yields a further quantity of less pure tin, which must be further treated.

Boiling.—The metal from the liquation furnace runs into the "refining kettle" in front. This is an iron pot about 4 feet 6 inches in diameter, heated by its own fire. Above the kettle is a lever attachment, by which logs of green wood can be held down in the molten metal.

The steam and gases disengaged by the action of heat agitate the metal and expose it to the air, and although tin is more readily oxidized than copper, bismuth, antimony, or lead, a scum of dross, consisting of iron, sulphur, arsenic, etc. forms on the top and is removed from time to time.

This is continued for from 1 to 8 hours, according to purity and the quality of tin required. For grain tin it is prolonged. This process is less an oxidation process than the effect of cooling at the surface, of metals less fusible than tin, which collect and form the scum. *Tossing* consists of lifting the metal from the pot in ladles and pouring it back again from a height of several feet. This is sometimes done instead of boiling.

For "common" tin the metal is ladled into the moulds, usually of granite, while boiling. For grain tin the metal after boiling is allowed to stand. The impurities yet remaining subside, and the upper purer layers are devoted to this purpose. The lowest layers require liquating again. Grain and refined tin are made from purer ores than common tin.

The purity of tin is tested by casting a small ingot in a stone mould. When pure, the edges are well rounded, and the surface remains brilliant when cold. Frosting of the surface on solidifying is an indication of impurity.

Manufacture of Tin Plate.

The principal uses of tin are the manufacture of alloys (see p. 265) for making tin plate, tinning cooking utensils, and for tinfoil.

Tin plates are sheets of iron coated with tin. The metal readily alloys with iron, and when heated somewhat above its melting-point, will adhere to a clean iron surface, forming at the surface of contact an alloy with the iron, to which a thin film of tin adheres. The adhesion of the film of tin depends on the homogeneity and purity of the iron employed, soft pure metal being most readily tinned. Such plates are also best suited for the use of the tinman, being more readily bent and worked.

The plates are rolled from "tin bars." These are about 6 inches wide and $\frac{3}{8}$ of an inch thick. They are cropped into 15-inch lengths, reheated, and rolled square. They are again reheated and rolled to about four times the length by chilled rolls. The plate is doubled on itself, reheated, and again rolled, again doubled, reheated, and rolled, and so on. The compound sheet passes through the mill as one sheet, sometimes as many as 32 thicknesses being rolled together. The sheets are cut to the required size and separated. A little coal-dust is sometimes sprinkled over the plates to prevent them sticking together, and the reheating is carefully managed to avoid overheating and consequent sticking together.

Formerly iron of special quality, smelted and refined with charcoal, was employed, but open-hearth steel is now generally used.

Preparation of Plates.—(1) The sheets of iron (black plates) are carefully annealed at a red heat. This is often dispensed with.

(2) They are then pickled in weak sulphuric acid at about 100° F., for about 20 minutes, well scoured with sand, and washed to remove scale (bright plates).

(3) The plates are annealed at a cherry-red heat in wrought-iron boxes from 10 to 12 hours, the air being excluded.

(4) The plates are cold rolled under chilled rolls to give an uniform, smooth surface.

(5) A second annealing of shorter duration and at a lower temperature than the first to remove the hardness produced by rolling is sometimes given.

(6) Another pickling in weaker acid than before, followed by scouring and washing to remove the thin film of oxide formed in the annealing processes.

The plates are then placed in water or lime water until required.

Tinning.—The sheets are first placed in a grease-pot containing melted tallow or palm-oil, somewhat strongly heated, and left till all the water has been driven off and the plates are uniformly heated and coated with grease.

From the "grease-pot" they pass into a hot bath of molten tin (tinman's pot), covered with grease or with zinc chloride, and strongly heated. Here the alloy on the surface is produced. It next passes to the "wash-pot," which is divided into two compartments, and contains tin, but at a lower temperature. In the first compartment the coating of tin is rendered uniform. The plates are lifted separately, and the surface brushed over with a hemp mop, and examined by the workman. If satisfactory, the plate is dipped rapidly into the second compartment, which contains pure tin, to remove the marks of the brush. It is then transferred to a grease-pot, where it passes through a pair of rolls, which squeeze off the excess of tin and improve the surface. The plates are then cleaned from grease in bran, rubbed with chamois leather or woolly sheepskin, and examined, faulty plates being rejected.

Formerly the plates, after tinning, were allowed to drain in a hot grease-pot, and the wire of tin which formed removed by immersion in a "list" pot, having about a quarter of an inch of molten tin at the bottom, in which the wire melted off.

Machinery has to some extent replaced hand labour for immersing the plates, particularly for large sizes and inferior qualities. The plates are carried through the several baths in succession by an arrangement of rolls and endless chain belts. *Terne* plate is an inferior quality coated with a lead-tin alloy.

Tinning Copper Articles.—The surface is first carefully cleaned, and the metal heated somewhat above the melting-point of tin, a little powdered resin or ammonium chloride is dusted over, and molten tin is then wiped over the surface with tow. A quarter of an ounce of tin will cover 2 square feet of surface, giving a durable coating.

Brass pins are tinned by boiling them with cream of tartar,

alum, salt, and granulated tin, in water. The tin is slowly dissolved by the liquor, and the zinc in the brass precipitates it on the surface.

Alloys of Tin (see p. 265).

CHAPTER XVIII.

ZINC.

THIS metal is commonly known as "spelter."[1] It has a bluish-white colour and high lustre. The brightness of the fracture is dimmed by impurity, notably by iron. Commercial zinc is highly crystalline, hard, and brittle. When pure, however, the metal is malleable, and ordinary zinc becomes sufficiently malleable to be rolled into sheets when heated to 120° to 150° C. If heated over 200°, the metal is more brittle than when cold, and can be powdered. It is harder than tin and softer than copper. It has a specific gravity of 7·1 when cast, but this may be increased to 7·2 or 7·3 by rolling. It fuses at 412° C., and is very fluid. It contracts very little in solidifying, and hence is suitable for castings. The nature of the casting is influenced by the temperature of pouring. If poured very hot, the castings are crystalline; but if near the melting-point, they are more granular. Zinc boils at the melting-point of silver, 950° C., and the vapour burns in air with a bluish-white and very brilliant flame, forming oxide of zinc (ZnO). The tenacity of cast zinc is 1·25 tons. After rolling and annealing zinc has a tenacity of 7 to 8 tons, and wire, 10 tons per square inch. The elasticity is high. Rolled zinc retains in some measure its malleability, and the hardness induced is removed by annealing at a *low* temperature. Zinc was formerly solely used for brass-making, the fact of its being malleable when slightly heated having been discovered early in the present century. The first rolling-mills were erected in Birmingham.

A little lead (under 1 per cent.) added to zinc intended for rolling is of advantage, but this renders it unsuitable for making strong brass.

[1] Brazing spelter is an alloy of equal parts of copper and zinc.

Chemical Properties.—Above its boiling-point zinc burns to ZnO—"philosopher's wool." Thus produced, it is in a light, feathery condition, hence the name. This oxide is white, non-volatile, and infusible at furnace temperatures, but becomes yellow when heated, and at very high temperatures agglutinates. It combines with silica and forms a very refractory silicate. It is reduced by carbon monoxide, carbon, hydrogen, and iron at temperatures above its boiling-point. Like iron, zinc is oxidized by carbon dioxide and steam.

Zinc is little affected by ordinary atmospheric influences. On exposure to moist air it becomes coated with a film of oxide of zinc, which, being insoluble, protects the metal from further action. This property is applied in coating iron articles with the metal by dipping them in a bath of molten zinc, the process being known as *galvanizing*. Before dipping, they are pickled in dilute hydrochloric acid, to remove the scale, and afterwards scoured, if necessary, and washed. They are then introduced into the molten zinc, covered with sal-ammoniac, which acts as a flux. Tin and lead are sometimes added to the bath, to improve the appearance.

Electro—cold galvanizing—is being successfully introduced.

Zinc is superior to tin as a coating for iron for atmospheric work, as it is electro positive to iron, and if the iron is laid bare at any point, the electric conditions set up, result in the zinc being dissolved and the iron preserved. Tin, on the other hand, assists in the more rapid attack on the iron at the bare place, owing to its being electro negative. Zinc being, however, readily attacked by vegetable acids, and also by alkalies, it cannot be used in contact with those bodies, nor in canning meats, fruits, etc. In towns where sulphurous acid and other acid vapours exist in the atmosphere, both zinc and galvanized articles are readily attacked. Salts in solution cause the action of water on it to be more rapid. The purer the zinc the less rapidly this occurs.

Pure zinc is not acted on by water, and but slowly by H_2SO_4 and HCl, but readily dissolves in nitric acid.

Zinc precipitates gold, silver, copper, platinum, bismuth, antimony, lead, tin, mercury, and arsenic from solution.

It does not readily combine with sulphur, but the sulphide is obtained by heating the oxide with sulphur, and by projecting a mixture of zinc-powder and sulphur into a red-hot crucible, ZnS being formed. This sulphide is practically infusible, and on roasting behaves like copper and lead sulphides, sulphur dioxide is evolved, and sulphate and oxide of zinc formed. A higher temperature is required to decompose zinc sulphate than is required for silver sulphate. Zinc sulphide is sometimes removed from ores by roasting at a low heat, and drenching the roasted mass with water to dissolve out the soluble sulphate of zinc formed. Sulphide of zinc is reduced completely by carbon and iron at high temperature, the zinc being volatilized.

Zinc Ores.

Red Zinc Ore—Spartalite—zincite (ZnO)—is generally red, owing to the presence of manganese. It is found associated with franklinite at *Franklin, New Jersey.*

Calamine—carbonate of zinc—($ZnCO_3$)—varies in colour, from white to brown. The brown colour is due to oxide of iron. It is generally of an earthy character. Some specimens are like entwined reeds, hence the name. It occurs in Flint, Somerset, Mendip Hills, Alston Moor in Cumberland, Lead hills in Scotland, Tarnowitz, in Silesia, Rhine Provinces, and Belgium (Aix-la-Chapelle), and in Spain. It usually occurs in limestone rocks, and is associated with the silicate. The Silesian calamines contain as much as 8 per cent. of silicate, and carry from 6 to 45 per cent. of zinc. Blende, galena, and sulphate of lead often accompany calamine. Lead and iron are both objectionable in zinc ores, on account of the corrosion of the retorts at the high temperature employed, by the oxides of those metals. Calamines are carefully dressed to remove lead as completely as possible before smelting.

Electric Calamine—hydrated silicate of zinc—is found associated with the carbonate.

Blende—black jack—zinc sulphide (ZnS) is the commonest zinc ore. It varies in colour from pale yellow to black, and has a resinous lustre. Pure ZnS is white, and the dark colour of blende is due to iron and other impurities. It is generally dark-coloured and crystalline. It commonly occurs associated with galena and pyrites, in limestone and other rocks. It is separated by careful dressing. It occurs in North Wales, Derbyshire, Isle of Man, Cumberland, Cornwall, Freiberg, United States, Russia, and many other localities.

Extraction of Zinc.

Zinc is extracted, in the treatment of simple ores, by the reduction of the oxide with carbon or carbonaceous matters, at a temperature above the boiling-point, so that the reduced metal is vapourized. The reduction is conducted in closed retorts, and the zinc vapour is led into condensers outside the furnace, and condensed. The discovery was made by Henckel, in 1721.

$$ZnO + C = Zn + CO$$

All ores are roasted to convert them into oxides, for although *carbonate* of zinc would readily reduce without calcining, the conversion of the CO_2 expelled from it into CO, would entail a large consumption of carbon, and a large increase in the volume of gas escaping from the condensers, with consequent greater loss of zinc.

$$ZnCO_3 = ZnO + CO_2$$
$$CO_2 + C = 2CO$$

Calamines are readily calcined on the bed of a reverberatory furnace, or often by the waste heat from the smelting furnaces. In Silesia, the small ore is treated in reverberatory furnaces, and the lump ore in kilns into which it is charged at the top with a little coal, and withdrawn at the bottom. Care is taken to keep the temperature too low to reduce and volatilize the zinc.

Blendes are usually calcined in long-bedded furnaces, with depositing flues, and the SO_2 is sometimes used for vitriol manufacture. At Oberhausen, it is used in manufacture of anhydrous sulphurous acid.

Note.—It is exceedingly difficult to calcine blendes "sweet," but the little sulphur remaining does not interfere with the reduction of the zinc; it is present as sulphate, which is decomposed by the higher temperature of reduction, and yields ZnO, which is reduced.

$$ZnSO_4 = ZnO + SO_2 + O$$

The reduction of the calcined ore is effected in closed vessels, crucibles or retorts, connected with suitable condensers. It is important (1) that the condensers should be of ample size, (2) that the exit for the gases should be contracted so as to prevent the oxidation of the zinc distilled off by the entrance of air into the condenser; (3) that the gases in the retorts and condenser should contain as little CO_2 as possible, or oxidation of the reduced metal will occur. To this end a high temperature and excess of carbon are necessary.

English Process.—In this process, introduced by Champion, at Bristol, early in the last century, the zinc ore (calamine) was mixed with carbon, and heated in large fire-clay crucibles, 4 feet high and 2½ wide at the top, having a hole in the bottom, to which a sheet-iron pipe, 6 inches in diameter, passing through the bottom of the furnace into a vault below, was applied. The lids of the pots were cemented on. The zinc vapour descended and was condensed in the tube—"Distillation per descensum." The method is wasteful, and has been entirely abandoned.

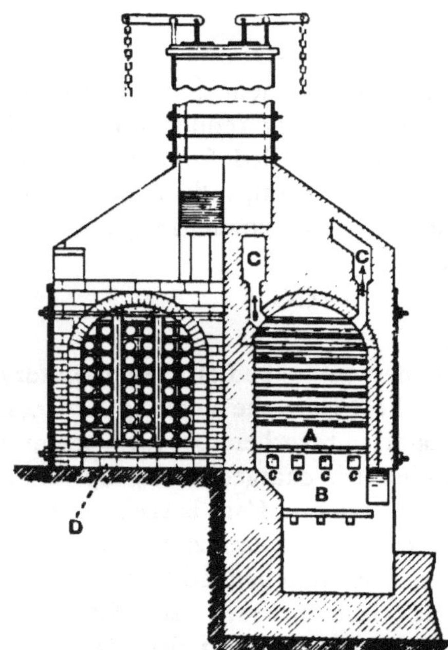

FIG. 89.—Belgian Furnace.

The Carinthian Process was somewhat similar, but fire-clay tubes were employed instead of crucibles. The zinc condensed in the lower part of the tube which projected through the furnace bottom. The process is not now employed.

The Belgian Process was introduced in 1810. It is conducted in cylindrical or elliptical retorts closed at one end and open at the other. They are about 39 inches long, 8 inches in diameter, and are supported at the ends, resting on the back and front walls of the furnace. They are somewhat inclined to the front, the open end being the lower, and are arranged in tiers one above the other.

Figs. 89, 90 show the arrangement. A (Fig. 89) is the furnace chamber—the back wall being vertical, with projecting ledges on which the ends of the cylinders rest. B is the fire-place; *c* the flues. The front of the chamber is closed by a cast-iron frame, D, protected on the inner side with fire-brick. Each of the compartments of the frame holds two retorts, which lie with their mouths resting on the framework. Each furnace holds from 40 to 80 of such crucibles, and they are built in blocks of four, back to back, with a common stack. The metal is condensed in fire-clay receivers of the form shown in Fig. 91; one of which is adapted to the mouth of each retort. A sheet-iron cone with a small aperture fits on this to condense fume. Before their introduction into the furnace, the retorts are carefully heated to redness. The spaces in the front frame are stopped with clay (cement of ground pots and raw clay), and the temperature gradually raised during 12 hours, the mouths of the crucibles being loosely stopped with clay plugs.

FIG. 90.—Belgian furnace.

FIG. 91.—Condenser with fume condenser attached.

The charge is a mixture of calcined calamine or blende with carbonaceous matters—anthracite or other non-caking coal, coke-dust, etc., as finely divided as possible, and slightly moistened. It is introduced by means of a scoop with a long handle. Each crucible receives a charge of from 13 to 27 lbs., the lower ones, being most strongly heated, receive the heavier charges.

After charging, the condensers are adapted (resting on a brick) and luted round. As soon as distillation commences, the fume-condensers are put on, a wet rag being often applied at the joint to prevent escape. The operation is judged by the flame of carbon monoxide which is ignited, and burns at the small opening at the mouth of the fume-condenser. At first the fumes are brown, and result from the cadmium contained in the ore which distils off first. This is succeeded by the greenish-white characteristic zinc flame, with white fumes, which continues as long as the operation lasts. The metal is scraped out of the condensers, which are hot enough to keep it molten, at intervals; the fume condenser is removed for this purpose. The distillation occupies about 12 hours. When completed, the residues are raked out of the retorts into the pit below, and recharging commences.

The yield of zinc by this process from ore containing 50 per cent. varies from 30 to 40 per cent. About one-half the remainder is recovered from the residue (fume, etc., caught in the fume-condenser), and the rest is lost, mainly as vapour, partly due to imperfect condensation and partly to cracked retorts, etc.

The fume is of course returned to the retorts for reduction.

The zinc is received from the condensers by a large iron ladle, and, after skimming, is cast into ingots. Gas firing has now become common.

Silesian Process.—The retorts employed in this process are ∩-shaped, and are supported throughout their length, thus permitting of the employment of a more intense heat without collapsing. They are about 39 inches long, 8 wide, and 12 to 18 inches high. One end of the muffle is closed. At the other end an opening at the top is provided for adapting the condenser, and another below for introducing the charge. This is, when working, closed by a flat fire-clay stopping.

The furnace is shown in Figs. 92, 93. It is divided into a series of bays on either side of the fireplace, each capable of containing two muffles, which rest on the bed, and are thus heated only on the crown and sides. The roof is dome-shaped. The ends of the retorts project through the side-walls of the furnace into a small outer chamber. From 12 to 32 muffles are contained in a furnace. The condenser is shown in Fig. 93, the metal flowing into a receptacle at *b*. An opening, *q*, at the bend, which is covered by a plate luted on during the distillation, is provided for removing obstructions.

This form of condenser has been largely superseded by that shown in Fig. 91, capped by fume-condensers, as in the Belgian process.

The operation lasts 24 hours, and the charge varies from 200 to 500 lbs. per muffle. The muffles last about 4 or 5

FIG. 92.

weeks. Cracks are stopped with a wash ot clay applied by a mop.

The loss of zinc in this process is somewhat greater than in the Belgian (though the residues contain less zinc), owing to greater leakage by cracking of the retorts.

Furnaces with a double row of muffles, one above the other, are employed at Freiberg and elsewhere, 64 muffles being sometimes used in a furnace of this kind.

Gas firing is now superseding the use of solid fuel. The type of furnace used is one in which the producer is attached to the furnace, the gas passing directly into the furnace chamber, or into a small combustion chamber, prior to entering the furnace. The air supplied for its combustion is heated by circulating through flues under the bed, heated by the waste heat from the furnace. Chambers for the calcination of calamine and the preliminary heating of muffles are also some-

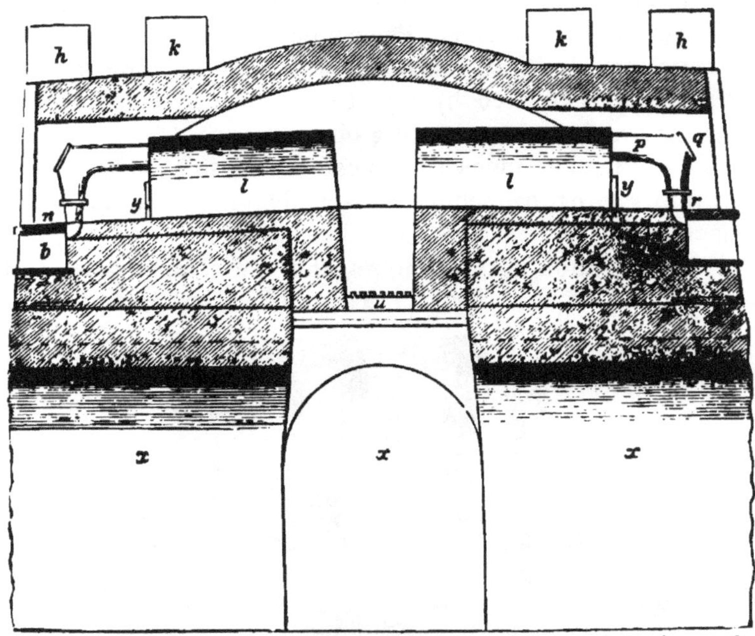

FIG. 93.—*l*, muffles; *y*, charging door; *p*, condenser; *b*, receptacle for zinc; *n*, fire-place; *x*, vaults.

times arranged to be heated by waste gases from the furnace.

Furnaces in which both Belgian and Silesian retorts are employed are also in use. The Silesian muffles occupy the lower part of the furnace, and the Belgian retorts are placed above. The more refractory ores are treated in the Silesian muffles.

Crude zinc contains iron, sulphur, arsenic, cadmium, and lead. The lead is volatilized and carried over in the distillation, as much as 4 per cent. being sometimes present. This is removed by fusion in a reverberatory furnace having a well at the lowest point of its inclined bed. The lead separates out and falls to the bottom of the well, and the zinc is ladled out. It still retains a little lead. Redistilled zinc contains about 0·2 per cent. Zinc for rolling is thus treated, as more than 1 per cent. seriously impairs its malleability.

Treatment of Zinc Fume.—The zinc fume which collects in the fume-condenser consists of oxide of zinc and finely divided metal. It is returned to the retorts and redistilled with carbon, or treated by the Montefiore process.

In this process the zinc-dust and oxide are placed in upright clay tubes, the ends of which pass through the bottom of the furnace. They are heated to 500° or more, and the finely divided zinc caused to run together by compression with a clay piston attached to an iron bar introduced into the tube. In from 2 to 3 hours the clay stopping at the bottom of the tube is pierced, and the collected metal flows out. The contents of the tube are again stirred up and compressed, and a second flow of metal obtained. As much as 80 per cent. of the zinc in the fume is thus obtained.

Blast-furnace Methods.—Proposals have been made for the reduction and volatilization of zinc in blast furnaces, the metal being obtained by arranging a series of condensing flues. The large volume of escaping gases which would render its complete deposition almost impossible, the oxidation by air and carbonic acid gas in the escaping gases—which are, of course, mixed with the zinc vapour—and many other reasons render this matter one of the greatest difficulty. In the most feasible methods the gases from the furnace are led through heated chambers or towers containing coke at high temperature, whereby CO_2 is reduced to CO, and any zinc oxidized again reduced. The metal is then obtained in condensing pipes kept hot enough to melt the zinc.

WET METHODS OF EXTRACTION.

Many wet methods have been proposed for the treatment of complex ores containing large amounts of iron, copper, and lead, which cannot be roasted and treated in the ordinary manner, owing to rapid corrosion of the retorts. The zinc, having been obtained in a soluble form, is precipitated as oxidé by lime, and reduced, or the solution electrolised.

CHAPTER XIX.

NICKEL.

THIS white, hard metal is used to a considerable extent, on account of its resistance to atmospheric action and its whitening effect on copper in the manufacture of German silver. It is also alloyed with steel. It is malleable, ductile, tenacious, and weldable; fuses at about the same temperature as iron, and is affected in the same way by the same impurities as that metal. Its specific gravity is 8·8, and, like iron, it is magnetic. It oxidizes when strongly heated, and combines readily with sulphur and arsenic, and is soluble in acids. It is found with iron in meteorites. In nature it occurs principally in combination with arsenic and sulphur, or as a hydrated silicate.

Arsenical ores are concentrated much as in copper smelting for the production of a *speiss*.

The process of extraction from speiss is essentially a chemical process, the nickel being obtained as oxide. This is mixed with lampblack and oil, the mixture compressed and strongly heated. The oxide is reduced by the carbon.

The silicate is run down in small cupolas with gypsum (sulphate of lime) or alkali waste (calcium sulphide); a matte of nickel and iron sulphide is obtained. The iron is removed as in copper smelting, and the pure sulphide obtained is roasted to oxide and reduced as before.

The formation of a compound of nickel with carbon monoxide ($Ni(CO)_4$), when CO is passed over freshly reduced oxide of nickel, is the basis of a new process for obtaining the metal. The nickel carbonyl is decomposed by passing its vapour through strongly heated tubes, when the metal is deposited on the sides.

Nickel is rendered malleable by the addition of small quantities of magnesium or manganese.

COBALT.

The principal uses of this metal are for colouring glass and glazes for earthenware. Its oxide gives a fine blue colour to glass. Its use in the metallic state is very limited. It is harder and more tenacious than iron, and is used to some extent for electroplating as superior "nickel" plate, and in alloy, for increasing the elasticity of bronzes. Its properties are similar to those of nickel and iron.

MANGANESE.

This metal has no application in the arts, except in alloy with other metals. It is a hard metal, and takes a fine polish. Its colour is white. It oxidizes quickly in moist air, and is dissolved by acids. Its affinity for oxygen is so great that the oxide is not reduced to metal when heated in hydrogen or carbon monoxide, only manganese protoxide being produced. The oxide is reduced when heated with carbon. The metal is also produced by the reduction of the chloride, in admixture with potassium chloride, by metallic magnesium or sodium in crucibles. It readily takes up carbon and silicon.

Rich alloys with iron, for use in steel making, are produced in the blast-furnace by smelting ores containing oxide of manganese, such as the Spanish manganiferous hematites.

It is also used in the manufacture of bronzes.

Chromium.

This metal is only used in alloy, generally in steel, on which it confers increased elasticity and hardness. Pure chromium is more infusible than platinum, and is as hard as emery. It is permanent in air, and may be heated to redness without oxidation.

The metal is obtained by reduction of the oxide at high temperature with carbon in lime crucibles, by electrolysis of the double chloride of chromium and ammonium, or by fusion of the sesqui-chloride with zinc or magnesium, the excess of zinc being afterwards removed by acid. Chromium is unattacked by nitric acid, but dissolves in sulphuric and hydrochloric acids.

Magnesium.

This is a brilliant silver-white metal, which, however, rapidly tarnishes in moist air. Its specific gravity is only 1·74. It is highly tenacious, about 14·5 tons per square inch. It fuses at about 800° C., and at a high temperature can be vapourized and distilled, like zinc. It burns in air with a brilliant white light, and is used for photographic purposes. Heated to 450° C. it can be worked, rolled, and pressed readily, giving forms of great exactness and sharpness.

It is malleable, but not ductile, except at elevated temperatures. Magnesium wire is made by *squirting* the metal, in a heated state, through holes in a steel plate. The ribbon is made by flattening the wire in heated rolls:

Minerals containing magnesium are abundant: magnesite, the carbonate ($MgCO_3$); dolomite ($CaCO_3MgCO_3$); carnallite ($MgCl_2.KCl.6H_2O$), kainit ($MgSO_4.KCl.6H_2O$) and Kieserite ($MgSO_4H_2O$) occur at Stassfurth.

The metal is prepared by decomposing a mixture of magnesium chloride mixed with sodium or potassium chloride, by metallic sodium, equal to $\frac{1}{5}$ or $\frac{1}{6}$ of its weight, in iron crucibles heated to redness. The resulting chloride of sodium is dissolved in water, and the magnesium purified by distillation in a wrought-iron still, provided with a lid secured by a screw. The still is connected by a tube which passes nearly to the cover, with an iron condenser beneath. The air is removed by a current of coal gas, before heating up the still. The metal is remelted and cast into ingots.

The metal may also be produced by electrolysis of the fused chlorides. (See Aluminium.)

Aluminium.

This metal, although fairly hard, is characterized by its extreme lightness. Its specific gravity when cast is only 2·56, which on rolling is increased to 2·68. It is highly malleable and ductile. Its tenacity is about 17 tons per square inch, and its elasticity is about equal to that of silver. At about 700° C. it melts, and contracts on solidifying.

In mass, it is unalterable in dry or moist air, at any temperature, but when finely divided, takes fire and burns on heating, forming the oxide Al_2O_3, which is not reducible by carbon at furnace temperatures.

It was formerly obtained by decomposing the double chloride of sodium and aluminium with metallic sodium. In the methods now followed for its extraction, the melted fluoride, or oxide, is decomposed by an electric current, the metal being liberated, as in the Cowles, Hall, and Heroult processes.

PLATINUM

Is a silvery or tin-white metal, almost as hard as copper. It is exceedingly malleable and ductile, being only inferior to gold and silver in these respects. Its specific gravity is 21·5. It is only fusible at the highest temperatures, *e.g.* in the oxy-hydrogen blow-pipe flame. It occludes oxygen like silver when molten. At a red heat it occludes nearly 4 times its volume of hydrogen. Its expansion by heat is 0·0000264 per degree, and is nearly equal to that of glass, 0·0000258. Wires can, therefore, be fused into glass without risk of breaking away—a point of great importance in the manufacture of electric lamps, etc. It welds when strongly heated.

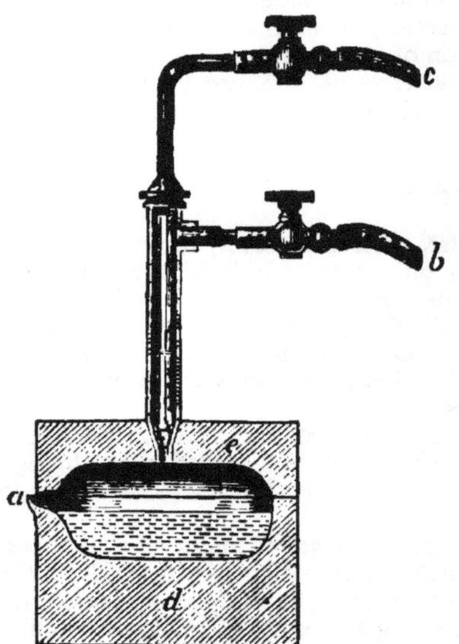

FIG. 94.—Lime furnace for fusion of platinum.

Owing to its not being readily attacked by acids or chemical reagents, it is largely used for making chemical vessels, such as crucibles and dishes, stills for concentrating vitriol, parting, etc.

It occurs native, in grains in alluvial deposits, and associated with the rare metals, rhodium, osmium, iridium, ruthenium, rubidium. After chemical treatment, it is finally obtained as the double chloride of ammonium and platinum. This is decomposed by heat, leaving finely divided, spongy platinum, which is fused by an oxy-hydrogen jet in a small furnace made of blocks of lime (Fig. 94); or the ore is smelted with galena, and the lead obtained cupelled.

ANTIMONY.

This metal has a bluish-white colour, and is highly crystalline and brittle. The surface exhibits fern-like markings. It has a specific gravity of 6·7 to 6·8, melts at about 450° C., and slowly volatilizes at a white heat. It expands slightly on solidifying, a property which it imparts to its alloys.

Its principal use is for hardening alloys of lead and tin, and it forms a constituent of type, stereotype, and Britannia metals.

Antimony occurs in nature principally as the sulphide—*stibnite* (Sb_2S_3)—from which it is obtained by heating with iron, in crucibles.

$$Sb_2S_3 + 3Fe = 3FeS + Sb_2$$

The crude metal is subsequently refined. The excess of iron is removed by heating the crude metal with more stibnite and fluxes, and "starred."

Bismuth.

This is a highly crystalline, brittle metal, of a white colour with a tinge of pink. Its specific gravity is 9·82. It melts at 268° C., and volatilizes at high temperatures. The vapour burns with a bluish flame. It expands on solidifying.

Its principal use is for adding to alloys of lead and tin—whose melting-point it lowers—for making "quick" solders for pewter, and fusible alloys (see p. 269).

It occurs *native* and as *sulphide*. The metal is simply liquated out of the ores in which it is native, and the sulphide is decomposed by iron, sodium carbonate being used as a flux. A considerable quantity of bismuth is extracted from the cupels used for cupelling rich silver lead alloys.

Cadmium.

Cadmium is closely allied to zinc, with which metal it is generally found associated. It is more volatile than zinc, and in the distillation of that metal comes over first. The vapour burns in air, giving brown fumes of cadmium oxide, CdO.

CHAPTER XX.

ALLOYS.

METALS are frequently mixed with each other for the purpose of modifying their properties in order to fit them for special applications. The principal objects are: (1) to harden; (2) to increase the strength, toughness, elasticity, or power of elongation; (3) to facilitate the production of sound and workable castings; (4) to lower the melting-point; (5) to modify the colour or structure; (6) to resist corrosion.

Thus gold is hardened for coinage and other purposes by the addition of copper and silver and occasionally zinc and other metals. Silver by copper, and so on. Copper is hardened by zinc, and its colour altered to yellow shades, in the various forms of brass. In gun-metal its strength[1] is increased by the addition of tin; its elasticity, strength, and power of elongation by nickel; while addition of zinc, etc., increases the soundness of the castings obtained.

Speaking generally, the alloying of one metal with another lowers the melting-point of the *less* fusible, and sometimes reduces it below that of the *more* fusible constituent.

The introduction of zinc and aluminium into copper to produce imitation gold alloys, and of nickel into brass to produce "nickel" silver alloys, are instances of modified colour.

The following list gives the metals in the order in which they affect

[1] In the cast state.

the colour of the alloy into which they enter, each metal producing a greater effect than that following it :—

1. Tin	4. Manganese	7. Zinc	10. Silver
2. Nickel	5. Iron	8. Lead	11. Gold
3. Aluminium	6. Copper	9. Platinum	

Thus an alloy of 1 part tin and 2 parts copper is white, but nearly 2 parts of zinc must be added to 1 of copper to whiten it. Most metals alloy together when melted, but many have a tendency to separate while cooling, owing to the difference in specific gravities. In the production of castings where this tendency is manifested, the alloy should be poured at as low a temperature as possible while ensuring the filling of the mould. The specific gravity of alloys often differs somewhat from that of the mean of the constituents, being sometimes above and sometimes below.

Heat is frequently evolved by the combination of metals. The tendency of metals to alloy with each other varies greatly ; thus, copper and zinc alloy well in all proportions ; of the copper-tin alloys, those represented by the formulæ Cu_3Sn, Cu_4Sn, Cu_7Sn, are the only ones that show no tendency to liquate, while the copper-lead alloys can be almost completely separated by liquation (see Silver, p. 228) ; similarly lead and zinc do not alloy (p. 198). The liquation of an alloy causes it to solidify piecemeal. Definite alloys separate at different temperatures throughout the mass, before complete solidification of the whole takes place. This can often be made apparent by etching the surface with acid.

The purity of the metals employed is of great importance, since minute quantities of impurity frequently exert a marked influence on the properties, 0·2 per cent. of bismuth in copper, used for alloying with gold for coinage, destroys the malleability to such an extent as to unfit it for that purpose.[1]

Production of Alloys.—(1) by fusing the metals together, or mixing them in the molten state ; (2) by compression of the finely divided metals (p. 177) ; (3) by electro deposition.

In making alloys, the metals, if not volatile and their fusing-points not too widely apart, may be melted together ; but if, as in the case of the copper-tin alloys, one metal is much more readily fusible, it is best added after fusion of the other has been effected.

When one of the metals is volatile, as in the copper-zinc alloys, the volatile metal should be added at as low a temperature as possible after fusion of the copper, in successive portions, each being kept below the surface till melted. In this way it is taken up by the copper as it melts, and is less readily volatilized. The first portions added lower the melting-point, and also cool the mass by absorption of heat in melting. Less loss of zinc occurs in this way. The mixture should be stirred. Oxidation should be prevented during melting by a covering of coke or other carbonaceous body.

Copper-Zinc Alloys.—These are commonly known as brass, but alloys containing tin also, are often thus designated. The zinc hardens the copper, causes it to cast sounder, and diminishes the toughness so as to permit of more easy working.

[1] Roberts-Austen, *S. and A. Jour.*, 1888.

The alloys are strong, and many of them are malleable. Lead diminishes the strength. Brass seldom consists merely of copper and zinc; iron, lead, etc., are often added for special purposes.

TABLE OF COPPER-ZINC ALLOYS.

Copper.	Zinc.	Tin.	Iron.	Properties.	Description.
55–60	38–44		1·5–4	Strong as mild steel, highly elastic, less malleable than other alloys.	Aich, Delta, and Sterro metal.
83	17			Softer than most alloys.	Red brass, tenacity 14·5 tons.
72	28			Malleable, ductile; rolls well; bright yellow colour.	Best brass, Bristol sheet.
66·6	33·3			Casts and works well.	Ordinary English brass.
60	40			Rolls well hot; resists corrosion.	Muntz or yellow metal for sheathings.
50	50			Yellow, unsuitable for rolling and drawing.	Common brass and brazing spelter.
66–73	27–34			Yellow, suitable for rolling and wire-drawing; very malleable and ductile.	Pinwire brass.
80–84[1]	15–20			Highly malleable yellow alloys.	Dutch, Bath, or gilding metal. Oreide gold.
75	20–25	0–5		Yellow, malleable; suitable for stamped work.	Mannheim, or Mosaic gold, Similor, Princes metal.
20–47	53–80			Brittle, but will bear slight pressure.	White brass; imitation platinum.

"Tombac" is a name given to alloys ranging from nearly pure copper to 30 per cent. of zinc.

ENGINEER'S BRASS.

This generally contains tin in addition to copper and zinc. Its composition varies from 75 to 90 per cent. copper, 2 to 16 per cent. of tin, and from 2 to 20 per cent. of zinc. It is tougher and stronger than ordinary brass.

[1] Tin sometimes added.

Table of Copper-Tin Alloys.

Copper.	Tin.	Zinc.	Lead.	Properties.	Description.
90	10			Very tough, finely granular, yellowish grey fracture, tenacious (18 tons).	Gun metal.
75–80	20–25			Hard, sonorous, brittle, homogeneous, granular.	Bell metal.
95	4	1			Coinage bronze.
82–92	2–6	3–8	0–3		Statuary bronze.
66·6	33·3			Hard, brittle, silver white, conchoidal fracture, takes a high polish, and is used for reflectors, etc.	Speculum metal; zinc, nickel, silver, and arsenic are sometimes added.

Copper-Antimony Alloys.—These metals alloy well. The alloy of equal parts of the two metals has a fine violet colour. It is hard, crystalline, and very brittle, and has no application in the arts. It is known as "Regulus of Venus." Antimony is sometimes mixed with brass to resist action of acids.

Tin, Lead, Antimony, and Zinc Alloys.—These comprise the soft solders, type metals, stereotype metals, pewters, etc.

Tin.	Lead.	Zinc.	Antimony.	Properties.	Description.
11		1		Very malleable and white.	Spurious silver leaf.
50		50		Casts well, fairly hard.	Pattern alloy.
45	10	45		Casts well and works easily under graver.	For small ornaments.
3	1			Hard and tenacious.	Fine solder.
2	1			Lowest melting-point of series.	Fine solder.
1	1				Tinman's solder.
1	2			Like most others of the series becomes plastic before solidifying.	Plumber's metal.
75–94	0–8	Cu 1–9	5–25	White, rolls and works well.	Britannia metal, for spoons and plate.
	80		20	Expands on cooling.	Type metal.
20	60		20	Lower melting point. Bismuth often added to lower melting-point.	For small type and stereotype.

Fusible Metals and Alloys.

Tin, Lead, Bismuth Alloys, used for fusible plugs and "quick" solders for pewter, etc.

Tin.	Lead.	Bismuth.	Cadmium.	Melting-point.	Uses and Remarks.
20	30	50		197° F.	For fusible plugs, taking impressions of dies, etc.
12·5	25	50	12·5	150° F.	Expands on cooling.
58·8	29·4	11·8			Pewterer's solder has a lower melting-point than work to be soldered.

Gold, Silver, and Platinum Alloys.

Gold.	Silver.	Platinum.	Copper.	Zinc.	Description.
	92·5		7·5		English standard silver.
	90		10		French and German coinage.
	75		25		German silver plate.
	91·66		8·34		Indian rupee, Brazilian coin.
	94·5		5·5		Netherlands coin.
	66·6		22·2	11·1	Silver solder.
91·66			8·33		British, Turkish, Brazilian gold coin.
98·9			1·1		Hungarian ducat.
90			10		German, French, Italian, Belgian, Spanish, United States, Swiss, and Russian gold coins.
10	6		4		Gold solder.
	65–83	17–35			Dental alloys.

See also p. 209 and 244.

Aluminium and Manganese Bronzes.

Aluminium Bronze.—The proportion of aluminium alloyed with the copper varies from 1 to 10 per cent. The alloys are as strong as mild steel, highly malleable, elastic and ductile. The presence of other metals impairs its quality. An alloy containing 10 per cent. has a tensile strength of 40 to 45 tons per square inch.

Manganese Bronzes contain copper, manganese, zinc, and tin. Sometimes they are characterized by hardness, elasticity and strength combined with toughness and resistance to corrosion. They can be rolled and forged hot. An important application is for the propellers of steam-ships. It is also used for general engineering brass work. The manganese is generally introduced in the form of ferro-manganese.

Phosphor Bronze is a bronze containing a small proportion of phosphorus, introduced either as phosphor tin (obtained by dissolving phosphorus in molten tin; it contains up to 20 per cent. of phosphorus) or phosphor copper, after fusion of the ordinary ingredients. The tin varies from 4 to 10 per cent., and the phosphorus from 0·1 to 1. Where toughness and ductility are required, the phosphorus should not exceed 0·1. Metals containing more, increase in hardness and are used for valves, bushes, cog-wheels, etc. It should be cast at as low a temperature as possible.

Silicon bronze contains silicon. It is harder and stronger than ordinary bronze.

The beneficial effects of phosphorus and silicon are generally attributed to the powerful deoxidizing influence they exert on account of their affinity for oxygen.

Nickel Alloys.

Nickel.	Copper.	Zinc.	Iron.	Tin.	Description.
14–31·5	40–56	23–26	2·3 / 3·5	0–4	Arguizoid, Chinese white copper.
15	60	25			Common German silver.
21	56	23			Medium ,, ,,
25	50	25			Good ,, ,,
28·3	38·3	33·3			Best ,, ,,
20	80				Cupro-nickel.

These alloys are white in colour, tough, and malleable. For rolling, a little lead is often added. The last is used as a sheath for the bullets of rifles, being hard and very suitable for drawing.

Amalgams. (See Mercury.)

Amalgams of tin and mercury, and mercury with cadmium, are employed for filling teeth. These amalgams become plastic by pounding or kneading when slightly warmed as in the hand, but set hard without contraction. A copper amalgam was formerly much employed, but it requires stronger heat to make it plastic. The alloy on the backs of looking-glasses contains 80 per cent. of tin and 20 of mercury (see Mercury). Sodium amalgam is produced by adding sodium to mercury; the combination causes great evolution of heat. It is prepared in considerable quantities for export. For this purpose it is packed in lime, to prevent access of moisture or carbon dioxide, in metal-lined cases. The amalgam contains about 3 per cent. of sodium, and is tolerably hard and semi-crystalline.

Iron Alloys.

Nickel Steel.—Nickel steel is being largely used for armour-plates. Generally from 1·5 to 2 per cent. of nickel is present. It increases the toughness of the metal, and diminishes atmospheric action and the action of sea water.

"Harveyized," armour plates are nickel-steel plates, case hardened on the surface by heating in contact with animal charcoal after the manner of making blister steel.

Chrome Steel.—This usually contains about 1·5 per cent. of chromium. Its presence increases the tenacity and hardness without diminishing the toughness, while the metal welds readily. It is used in the manufacture of shell-cases.

Tungsten Steel.—Mushet's special steel is a self-hardening tool steel, containing up to 9 per cent. of tungsten. It is extremely hard and strong, breaks with a conchoidal fracture, which has a faintly yellowish or brownish tinge.

Molybdenum is being introduced for the same purpose, a smaller quantity producing similar results.

Aluminium is added to steel to produce sound castings. The "Mitis" castings owe their superiority to the presence of this metal.

Manganese Steel.—Manganese in excess produces great hardness. The alloy is tough, but almost unforgeable. It contains from 9 to 13 per cent. It is very fluid when molten, and casts soundly.

Iron and Zinc.—Iron dissolves in zinc, heated to nearly its boiling-point to a considerable extent. The iron in delta metal is introduced by saturating molten zinc with iron, and adding this alloy to the copper in sufficient amount. A hard alloy of zinc and iron forms in galvanizing pots, and zinc melted in iron vessels takes up iron.

INDEX

D

E

F

N

O

P

Q

PRINTED BY WILLIAM CLOWES AND SONS, LIMITED, LONDON AND BECCLES.

www.ingramcontent.com/pod-product-compliance
Lightning Source LLC
Chambersburg PA
CBHW021511210326
41599CB00012B/1217

* 9 7 8 3 3 3 7 2 7 5 9 7 6 *